SCIENTIFIC AND EDUCATIONAL JOURNAL OF THERAPEUTIC RIDING

2011

ISSN 1174-6238

ISBN 978-0-9845490-3-0

A PUBLICATION OF

THE FEDERATION OF HORSES IN EDUCATION AND THERAPY INTERNATIONAL A.I.S.B.L.

The Federation of Horses in Education and Therapy International A.I.S.B.L

SECRETARIAT

HETI
P.O.Box 134
Damariscotta, ME 04543
Tel: +44 207 1831 256
Email: office@HETIfederation.org
www.HETIfederation.org

ISSN 1174-6238

ISBN 978-0-9845490-3-0

Table of Contents

Editorial

Psychotherapeutic activity involving horses is a growth industry, derived from very different schools of thought and practice, about which I receive increasing demands for information. In addition, there are well intentioned and valid concerns about the professional location and standing of this type of therapy.

With this in mind I approached a number of organisations and individuals with the aim of showing different perspectives of psychotherapeutic work involving horses;, to stimulate debate, to encourage further professionalized practice, and thereby to help protect vulnerable patients/clients. The response was positive. Regrettably no response was received from Linda Kohanov and, for understandable reasons, the German schools which have been leaders in this field, were unable to contribute to this edition of the Journal.

The Journal begins and ends with different historical perspectives. Both are well grounded in extensive referencing which I trust will assist researchers and others to expand this area of knowledge.

Alison Selby's introductory paper sets the scene for three psychotherapeutic models:

- Lynn Thomas' concise portrait of the increasingly familiar EAGALA model established in 1999,
- Sven Forsling's little known, but highly successful, Frossarbo model, developed in the 1980's in Sweden to evaluate a different form of institutional care within the Swedish penal system, and
- Leigh Shambo's HEAL model focussing on bonding with the horse as treatment for Post-Traumatic Stress Disorder (PTSD) and other clinical disorders.

Leif Hallberg, drawing on her recent experience as chair of PATH Intl.'s EFP/L Task Force, gives another perspective in her elaboration of PATH Intl.'s role and method in contributing to sound practice and interface with relevant professional bodies.

In the final paper Corey Souza provides a thought provoking historically linked conclusion by locating the intersection of hippology and anthropology from early domestication to hippotherapy.

Two other research papers on aspects of fear and safety, while of more general applicability nonetheless, fit well with the psychological perspective of this Journal. Fear, as addressed by Anita Shkedi and the risk management issues in Rebecca Cook's study are fundamental to all work with horses. We hope to report further on Rebecca's findings as she continues to track incidents and injuries.

As in previous years, terminology continues to be an issue in need of resolution. Research (and editors) would be well served if organisations could put aside their differences and agree on international terminology.

Ann Kern-Godal
Department of Addiction Treatment (Youth)
Center for Substance Abuse and Addiction Treatment
Oslo University Hospital, Norway

The Federation of Horses in Education and Therapy International A.I.S.B.L. would like to acknowledge with gratitude the generous contribution of an anonymous sponsor from Belgium. This donation made possible the design, printing and distribution of this publication.

Article One

A Historical Perspective of Psychotherapy Involving Equines[1]

Alison Selby, MSSW, LMSW
Correspondence to:
Alison Selby, Child and Family Guidance Center
4031 Plano Pkwy, Plano, TX, 75093, USA
Phone: +1-214-673-6010
E-mail: aliby@verizon.net

Abstract

Equine-facilitated psychotherapy is explored in a historical context to enhance understanding of the treatment approach and as a basis for the continued construction of a theoretical framework. An overview of the European and North American history of horsemanship as therapy is presented, and the preeminent professional organizations are identified. Hippotherapy, therapeutic horsemanship, therapeutic riding, and other activities with horses are described as they relate to therapeutic techniques involving equines in the mental health setting. The importance of standardized terminology is emphasized to ensure proper usage, and to promote comprehension and agreement across settings. Some relevant theoretical constructs are suggested, which include animal-assisted therapy, biophilia, myth and metaphor, as well as theories of attachment, neurodevelopment, the therapeutic relationship and exercise; they are identified as important foundational elements of the evolving theoretical underpinnings of this unique form of treatment.

Keywords

Equine Therapy; Equine-Facilitated Psychotherapy; Equine-Assisted Psychotherapy; Equine-Assisted Activities; Therapeutic Horsemanship; Therapeutic Riding.

---

[1] This paper is excerpted from the introductory material of a systematic review of the literature conducted in fulfilment of the requirements for completion of the Master of Science in Social Work degree at The University of Texas at Arlington, May, 2009. The full systematic review has been accepted for publication in the journal Health Psychology.

Introduction

Horses are powerful, and their involvement in therapies for people with various disabilities and difficulties are not recent phenomena: History is rich with accounts of the extraordinary curative effects associated with interaction with equines. Examination of the historical context of the inclusion of horses in therapeutic interventions furthers understanding and is a useful foundation for the construction of a theoretical framework, which ultimately enhances the credibility and encourages further development of this unique treatment strategy.

European History

The origins of horsemanship as therapy can be traced back to the Classical era when Greek and Roman texts described the beneficial relationship between horses and people (Bracher, 2000; Butt, 1981/1998; Fosdick, 2003; Hallberg, 2008; MacKinnon, Noh, Laliberte, Lariviere, & Allan, 1995). Ancient treatises on medicine by Galen and Orebasius allude to the therapeutic effects of riding (Butt; DePauw, 1986; Fosdick; Haskin, Erdman, Bream & MacAvoy, 1974). Between 460-377 BC, Hippocrates included riding in a chapter on "natural exercise", and in 1569, Huronymus Merkurialis of Italy wrote *The Art of Gymnastics*, which discussed riding and its beneficial effects on the restoration and maintenance of health (American Hippotherapy Association, n.d; Fosdick; Hallberg). English physician Lord Thomas Sydenham wrote in 1670: "There is no better treatment for the body and the soul than many tours each week in the saddle, riding the horse." (Butt, 1981/1998, p.1; see also Hayden, 2005, p. 60); Sydenham was a strong proponent of the benefits of riding for ailments ranging from hysteria and hypochondria to gout (Sloan, 1987).

A more modern reference to the physical and emotional benefits of horseback riding can be found in 1875, when the French physician Chassaigne recommended riding as a treatment for the improvement of deficits in posture, balance and joint movement manifested in some neurological disorders, and also noted the attendant psychological benefits of horsemanship (Acampora, 2008; Bain, 1965; Biery, 1985; Mayberry, 1978). Soldiers returning to England with injuries sustained in World War I were treated with riding therapy at the Oxford Hospital (Fosdick, 2003; Scott, 2005). But in 1952 the therapeutic possibilities inherent in horse activities came to the fore emphatically when Lis Hartel of Denmark, disabled by the effects of polio, won a silver medal in dressage at the Helsinki Olympics (Biery; Engel, 1998; Gatty; Harpøth, 1970; Henricksen, 1971), the first Olympics in which women were allowed to compete in equestrian sports. Lis Hartel subsequently repeated her accomplishment in the 1956 Stockholm Games (Database Olympics, 2008; Politiken.DK, 2009); her success prompted recognition from both medical and equine professionals that an exciting form of rehabilitative therapy deserved greater attention. Soon after, therapeutic riding began to be used increasingly for physical rehabilitation, predominantly in England, Germany and Scandinavia, and subsequently in North America (Butt; Bieber, 1998; Fitzpatrick & Tebay, 1998; Hallberg, 2008; Mayberry; West, 1970; Wingate, 1982).

North American History

Although therapeutic horsemanship as a means of physical rehabilitation had been recognized in Europe for some time, the concept did not arrive on the North American continent until 1960 (Butt, 1981/1998; Bieber, 1998). At that time the foundation for the Community Association of Riding for the Disabled (CARD) was laid in Toronto, Canada, and in 1963 Dr. Elmer Butt initiated a "riding for the handicapped" program in Windsor, Ontario (Butt). Upon her return from England, where she had been introduced to the beneficial possibilities inherent in riding for people with disabilities, Maudie Hunter-Warfel established the National Foundation for Happy Horsemanship in Malvern, PA, the first organized association for therapeutic riding in the United States (Bieber, Griffith, 1993); she was instrumental in disseminating knowledge and safety practices to other interested parties. Concurrently, the Cheff Foundation in Augusta, Michigan, was establishing the groundwork to begin construction of the first purposefully built facility devoted to horseback riding for people with disabilities, which was completed in 1967 (Butt). Mary Woolverton, a social worker and visionary, is notable for her early work with amputees returning from Viet Nam. She not only offered veterans disabled in war new opportunities for freedom of movement astride a horse, but pioneered carriage driving for people with disabilities (Bieber).

NARHA/PATH INTL

In 1969 the North American Riding for the Handicapped Association (NARHA) was formed in Middleburg, Virginia, to promote equine-assisted activities for individuals with disabilities (Butt, 1981/1998; Bieber, 1998). To reflect a more modern "people-first philosophy", and to encompass the wide variety of equine-assisted activities engaged in by its many members, in November, 2008, NARHA dispensed with the acronym and officially adopted NARHA as the name of the organization (NARHA, n.d.). On June 20, 2011, NARHA became the Professional Association of Therapeutic Horsemanship International (PATH Intl.) in an effort at inclusion of the many equine-assisted activities and therapies that continue to develop (PATH Intl., 2011). PATH Intl. continues as the primary governing body for professionals in North America, and particularly the United States, in the field of therapeutic horsemanship activities, offering certification for individuals and facilities and publishing the journal Strides to disseminate information pertinent to the practice of therapies involving horses.

Equine-facilitated Mental Health Association (EFMHA).

It is becoming increasingly apparent that horsemanship is not only beneficial for those manifesting developmental or physical disabilities: Empirical literature is beginning to emerge in support of the psychosocial benefits of the horse-human relationship. In response to the growing awareness of the mental health benefits of specific targeted equine-assisted activities, mental health professionals interested in utilizing horses as a therapeutic modality formed the Equine Facilitated Mental Health Association (EFMHA) in 1996, under the NARHA umbrella (Moreau & McDaniel, 2000). Beginning in 2009, as efforts were underway to restructure and rename the organization, NARHA began the process of integrating EFMHA back into the larger organization, to "promote the provision of services to the whole person" (A. Alden, personal communication, 2009).

Equine-assisted Growth and Learning Association (EAGALA)

In July, 1999, parallel and unrelated to EFMHA/NARHA, a group of professionals in the mental health field joined to create the Equine Assisted Growth and Learning Association (EAGALA), which aims to boost acceptance of equine assisted psychotherapy (EAP) as a valid and effective experiential approach for utilization by professionals in the clinical mental health and human development fields (EAGALA, n.d.). EAGALA's mission is to provide "...education, standards, innovation, and support to professionals providing services in Equine Assisted Psychotherapy" [EAP] (2005). Apart from those involved with NARHA, the professionals who comprise EAGALA have developed the "EAGALA model", which utilizes the horse as "...a treatment modality to foster emotional growth and learning through groundwork and structured activities with horses" (EAGALA; Kersten, 1997; Thomas, 2001). Targeted specifically to practice in mental health settings, the EAGALA model is not based on teaching riding or horsemanship; instead, it has been developed as a set of brief, experiential psychotherapeutic techniques, and has been compared to the familiar ropes courses used extensively by many therapists and treatment facilities. Using the EAGALA model, therapists address a variety of mental health and development needs which may include issues surrounding depression, anxiety, trauma, behavior problems, eating disorders and substance abuse.

OTHER ORGANIZATIONS

Capitalizing on the widespread and intuitive appeal of activities involving horses, other organizations have begun to emerge which incorporate horses in various experiential activities designed to promote education, leadership skills, and general well-being. A few of the most well-known organizations are Adventures in Awareness, Epona Equestrian Services and the Equine Guided Education Association. Barbara Rector, a pioneer in the development of equine-facilitated experiential learning, is the director of Adventures in Awareness, where the mission is to "develop consciousness while expanding awareness" (Rector, n.d.). Epona Equestrian Services offers leadership training and organizational development programs (Kohanov, 2003). The Equine Guided Education Association (EGEA) provides experiential education and "...supports and promotes human growth, learning and development through horse as healer/teacher interactions" (EGEA, n.d.).

HIPPOTHERAPY AND THERAPEUTIC HORSEMANSHIP/RIDING

Frequently, the terms hippotherapy (HPOT) and therapeutic riding (TR) have been used interchangeably; however, there are critical conceptual differences between TR and HPOT. As Glasow (2007) emphasizes, the matter of semantics in this field is very delicate. TR is similar to any horseback riding in that it is a strenuous sport that involves risk; it is a physical activity that increases general health by stimulating the cardiovascular system and strengthening muscles in the same way as other sports, with the added benefit of the soothing mental and social effects of bonding with an animal (Engel, 1998). While it can be argued that all riding may be therapeutic, not all riding is therapy.

The term "hippotherapy" was derived by the Germans from the Greek word for horse, "hippos", compounded with the word "therapy", to mean "treatment with the help of a horse" (Glasow; Debuse, Chandler & Gibb, 2005). The use of the movement of the horse is the therapeutic tool, not the horse itself, and members of AHA (American Hippotherapy Association) are careful to denote the distinction. HPOT sessions are conducted by licensed physical therapists, occupational therapists, or speech and language pathologists specially trained in the principles of this treatment approach, and facilitated with a carefully trained team consisting of a horse specialist, horse leader and/or long-liner, and sidewalkers, in addition to the client and the horse (AHA; Meregillano, 2004).

HIPPOTHERAPY

Unlike therapeutic riding, HPOT refers to a passive form of riding in which the client benefits from, but does not control, the movement of the horse (Silkwood-Sherer & Warmbier, 2007). Riding skills are not taught, and often bareback pads are used instead of saddles so that the client can benefit not only from the movement, but also from the warmth, of the horse (Snider, Korner-Bitensky, Kammann, Warner, & Saleh, 2007). Clients may be positioned astride facing forward or backward, sit sideways, or lie prone or supine (Benjamin, 2000; Snider et al.). Ideally horses are long-lined, (or ground driven), rather than simply led, to ensure as much straightness and correct movement as possible. HPOT is not a distinct treatment strategy that is mutually exclusive: rather, it utilizes the movement of the horse with a variety of treatments such as the neurodevelopmental approach, sensory integration, motor learning, motor control, and psycholinguistics to address neuromusculoskeletal dysfunction.

It has been suggested that the movement of the horse closely simulates human ambulation (Benda, McGibbon & Grant, 2003; Bertoti, 1988; Biery, 1985; Engel, 1998; McPhail, 2006; Silkwood-Sherer & Warmbier, 2007; Snider, Korner-Bitensky, Kammann, Warner & Saleh, 2007). The goal of treatments using HPOT is the habilitation or rehabilitation of persons with specific medical deficits or dysfunction (Biery; Engel; Maregillano, 2004). HPOT is used to address specific problems using a defined treatment protocol provided under a physician's prescription; in the controlled environment, the therapist modifies the horse's movement and carefully grades sensory input, thereby establishing a foundation for improved neurologic function and sensory processing (Benjamin, 2000; Meregillano, 2004; Sterba, Rogers, France, & Vokes, 2002). Postural and motor responses are often the primary focus of the HPOT session; positive effects have been noted in motor coordination, muscle tone, postural alignment, stiffness/flexibility and strength (Benjamin, 2000). Changes have also been reported in respiratory, cognitive, sensory processing, balance, affective, arousal and speech/language systems as a consequence of postural and motor improvements (Benjamin, 2000).

American Hippotherapy Association (AHA).

As the horse began to be viewed as a valuable adjunct to physical therapy, first in the Scandinavian countries following Lis Hartel's success, and then in Germany, Switzerland, and Austria, physical therapists in the United States began to develop treatment plans incorporating the movement of the horse (Borzo, 2002). In 1987, a group of eighteen American and Canadian therapists traveled to Germany to study the dynamics and applications of HPOT, with the goal of developing a standardized curriculum: This led to the development of the American Hippotherapy Association (AHA) in 1992 and in 1993 AHA was approved as the first subsection of NARHA. Shortly thereafter, AHA established therapist registration and standards of practice and, in 1999, subsequent to the establishment of the American Hippotherapy Certification Board, the premier Hippotherapy Clinical Specialist examination was administered (AHA, n.d.). In 2004, AHA seceded from NARHA and was installed as an independent governing body for the accreditation of therapists endeavoring to employ this treatment approach.

Therapeutic Riding, Horsemanship, and Other Activities

Therapeutic horsemanship (TH) involves any of a number of active physiotherapeutic exercises on and around the horse (Biery, 1985; Engel, 1998), which may include unmounted activities with the horse and/or in the horse environment, such as grooming and stable management experiences. In therapeutic riding (TR), traditional riding skills are taught by a certified therapeutic riding instructor. With improvements made during the course of HPOT, clients can sometimes progress to TR, where they take an active role in the control of the horse's movement and direction.

Therapeutic carriage driving is an exciting aspect of the equine tradition which offers participants who may not be able to ride the opportunity to take part in alternative mounted activities (Griffith, 1992). Interactive vaulting engages participants in horsemanship activities and movements around, on, and off the horse or barrel and ultimately in performing gymnastic exercises on the moving horse (Biery, 1985; Bracher, 2000; Frewin & Gardiner, 2005; Kroeger, 1992; Schultz, Remick-Barlow & Robbins, 2007; Vidrine, Owen-Smith & Faulkner, 2002). Vaulting is most often carried out in a group format, and consequently the opportunities for social engagement and interaction are multiplied exponentially.

Equines and Mental Health

Involving horses in a therapy setting is different from the typical companion animal-human interaction in that horses are not predatory by nature as are dogs and cats, but are rather themselves animals that are preyed upon (Blake, 1975; Fredrickson, 2008; Irwin & Weber, 2005; Morris, 1988; Rashid, 2000; Scanlan, 1998); consequently, equines offer unique opportunities in the therapeutic process. Horses are highly social and some of the attributes that they bring to the therapeutic environment are cooperation, patience, willingness, receptiveness, and, after millennia of domestication, an orientation toward people (Ewing, MacDonald, Taylor & Bowers, 2007; Hayden, 2005; Karol, 2007; McDaniel, 1998; Morris; Vidrine, Owen-Smith, & Faulkner, 2002).

The dynamic interchange which occurs between clients and horses offers a dimension to clinical work which is not possible within the traditional confines of the office setting. Because horses are prey animals, their survival depends on their extreme sensitivity to the environment (Rashid, 2000; Scanlan, 1998). They are essentially living biofeedback providers because of their ability to respond to the emotions and internal states of those around them. Regardless of how much a person tries to disguise emotional states, horses ignore outward form and instead respond to inner substance (McDaniel, 1998; Roberts, Bradberry & Williams, 2004). The horse's demands are relatively simple and uncomplicated, and conflicts are brief and few (Morris, 1988; Scanlan). Psychosocial interventions involving equines concentrate on the social dynamics of equine behavior to help individuals recognize and solve their own problems while focusing on an external, sentient being, which helps to develop awareness of internal motivations and issues (Dell, Chalmers, Dell, Sauve, & MacKinnon, 2008).

Horses are direct and honest in exchanges, whereas humans confuse and change the rules of social interactions through verbal communication (Russell-Martin, 2006). The horse has been described as a "mirror" for human emotions and processes (Kersten, 1997; Thomas, 2001; 2002). Without self-constructed barriers, they help individuals learn to be congruent in their words and actions (Moreau & McDaniel, 2000; Schultz, Remick-Barlow & Robbins, 2007). If a client is struggling with boundary issues and moves too close to a horse, the horse will take care to protect its space. Because of their social nature, opportunities to teach social and relational skills abound in the equine environment, since that is their basic mode of survival. Establishing communication with a species other than our own helps to develop skills that can be applied to cross-cultural communication, so important in an ever-increasing global economy.

Terminology

In order to ensure agreement among professionals, it is essential to understand the definitions of key terms. Improper usage and lack of agreement of the terms used to describe different aspects of therapeutic activities involving horses has led to general confusion and is a major contributor to the difficulty in measuring outcomes (Glasow, 2007; Silkwood-Sherer & Warmbier, 2007). In addition, the rapid growth of a wide array of activities involving horses has contributed to the general misunderstanding and misapplication of terminology. Although major efforts are underway by the leading professional membership organizations to standardize terminology, the lack of a single unifying body makes this a difficult task.

Equine-assisted activities (EAA) is a broad term which describes any of a number of therapeutic or educational approaches. Equine-facilitated mental health (EFMH) describes the alliance between a licensed mental health professional and a professional horseperson, "...in which the experience and knowledge of both are used to foster mental health in their students and clients" (Moreau & McDaniel, 2000, p. 2; see also Kersten, 1997). Equine experiential learning (EEL) involves unmounted and/or mounted lessons conducted by a licensed horsemanship instructor (Moreau & McDaniel; Fredrickson, 1992; NARHA, n.d). Equine-facilitated and equine-assisted psychotherapy (EFP/EAP)[2] utilize mounted and unmounted sessions provided by a trained and licensed mental health professional and a trained, licensed riding instructor (Moreau & McDaniel).

Theoretical Underpinnings

Because EFP is just emerging as a viable adjunct to traditional psychotherapeutic techniques, the theoretical foundation of why and how it works are still in the early stages of formulation (Roberts, Bradberry & Williams, 2004). In addition to the lack of a standardized terminology in the broad field of EAA, the lack of a clear theoretical framework is problematic and contributes to a continuing sense of

---

[2] Controversy/confusion occurs about use of the terms EAP/EFP. They are used interchangeably by some organisations, such as EFMHA, while others, such as EAGALA, use one or other to name a specific type of therapeutic activity (Hallberg, 2008:287)

disorganization, although conceptualizations for a theoretical basis for the effectiveness of therapy involving equines are beginning to emerge.

Animal-Assisted Therapy

While comparatively little quantitative research has been published in peer-reviewed journals documenting the effectiveness of therapeutic interventions utilizing horses and other equines, there is a wealth of information available concerning the healing effects of therapy involving companion animals (Corson & Corson, 1979; see also Taylor, 2001). Animals have been used for therapeutic purposes in a variety of settings (Beck & Katcher, 1996; Chandler, 2005; Levinson, 1962; 1965; 1969a; 1969b; 1971); they embody qualities and characteristics that inspire and motivate people to participate in constructive activities in which they might not typically engage. Sociologist James Bossard (1944) received over 1000 responses to his article on the mental health benefits of dog ownership in the journal Mental Hygiene. Boris Levinson (1962) described the benefits of having an animal present during therapy sessions with some patients (see also Chandler; Netting, Wilson & New, 1987; Serpell, 2000), which he reported to have discovered accidentally when his dog Jingles elicited a positive response from a reportedly treatment-refractory child (Eggiman, 2006; see also Fine, 2000; Taylor, 2001). Interestingly, it is reported that Freud believed that dogs have the ability to judge a person's character (Serpell; Eggiman), and that his Chow dog, Jo Fi, was present at all of his therapy sessions, signaling the end of the treatment session by pawing at the door.

Bonding with companion animals has been shown to be a useful treatment alternative for people experiencing distress associated with loss, alienation, trauma and other forms of disequilibrium (Christian, 2005; Folse, Minder, Aycock & Santana, 1994; Granger & Granger, 2004; Hansen, Messinger, Baun & Megel, 1999; Haynes, 1991; Hines, 1983; Hines, 2003; Levinson, 1962; see also Fine, 2000; Eggiman, 2006; Kruger, Trachtenber & Serpell, 2004; Reichert, 1998; Yorke, 1997, 2008; Yorke, Adams & Coady, 2008); physiological benefits of bonding with companion animals have also been shown (Barker & Wolen, 2008; Batson, McCabe, Baun & Wilson, 1998; Beck & Katcher, 1996; Delta Society, 1996; Friedmann, Thomas and Eddy, 2000; Garrity & Stallones, 1998). Therapy with companion animals has been employed in schools (Katcher, 2002; Katcher & Wilkins, 1998; Kruger, Trachtenberg & Serpell, 2004); in institutions and prisons (Barker & Dawson, 1998; Cushing & Williams, 1995; Katcher, Beck & Levine, 1989; McVarish, 1995; Moneymaker & Strimple, 1991; Ormerod, 1998); with children (Grier, 1999; Reichert, 1998: Robin & ten Bensel, 1985), and adults (Holcomb, Jendro, Weber & Nahan, 1997).

Biophilia

As early as 700 BC, the Greek Hesiod recommended in his Works and Days the serene and structured life at the farm as an antidote to the stress of life in the city (Katcher & Wilkins, 1998). Frederick Law Olmstead, the designer of Central Park in New York, wrote about his intuitive conviction that natural settings could promote recovery from stress and increase mental functioning (Ulrich, 1993). Recommendations of a retreat to the country remained a common suggestion to remedy the ill effects of the stress inherent in living in urban environments until the 1970's (Biery, 1985; Netting, Wilson & New, 1987). It seems that throughout history, people have gone to considerable lengths to maintain contact with nature, part of the justification for planning urban parks and natural areas and preserving wilderness for public use.

In 1984, Harvard researcher E. O. Wilson offered the term "biophilia", which he conceived to describe "[T]he innately emotional affiliation of human beings to other living organisms" (Wilson, 2007, p. 249; see also Kellert, 1996; Kellert & Wilson, 1993). He suggested that as humans moved through the evolutionary process, a biologically-based attraction for nature and its life forms developed (Kellert). Numerous studies have demonstrated that inhabitants of industrialized societies prefer specific landscapes (Ulrich, 1993;

White & Heerwagen, 1998), namely open spaces and grassy meadows, of the type commonly found in typical equine-friendly environments.

Because clinical problems generally are conceptualized as involving difficulties in the social and not the natural environment, clinicians have not usually addressed clients' relationships to the natural environment (Besthorn & Saleeby, 2003). The resurgence of interest in nature-based therapies and an accumulating body of research shows that exposure to pleasant nature stimuli promotes positive emotional states and respite from stress (White & Heerwagen, 1998). In the popular book *Last Child in the Woods*, Richared Louv (2006) describes what he terms "nature-deficit disorder" to account for many of the ills in modern-day culture, particularly among children and youths. Despite this evidence, investigators have devoted little attention to the incorporation of natural environments into psychosocial treatments, given that our relationship to the environment is as much a part of our deep history as social behavior itself (Wilson, p. 13). Therapies that occur in nature, such as EFP, are a means of gaining an appreciation for, and a way to halt the destruction of, the natural element so important to our well-being.

Myth and Metaphor

The peculiarly human quest for meaning finds natural expression in the embodiment of animals. The development of language, myth, and thought appears to be greatly dependent on the use of nature symbols, particularly animals (Kellert, 1996): Their life force, which in ancient times manifested as the supernatural or divine in anthropomorphic form, is the same power that is tapped into today in animal- and equine-assisted therapies (Root, 2000). Animals, particularly horses, awaken a primal sense of freedom which seems to be missing from much of harried life in modern times. As language and culture expand humans look to animals as a principle source of myth and metaphor.

Literature abounds with animals portrayed in myths, fairy tales, dreams and nightmares (Robin & ten Bensel, 1985). Horses, and even the equipment used when incorporating horses in therapeutic sessions, often provide compelling metaphors for the difficulties clients may be experiencing and the recovery process (Christian, 2005); they provide immediate opportunities to construct or rewrite narratives which enhance well-being. Humans are prone to story-telling: Cultures have evolved by people sitting around campfires telling stories, constructing and reconstructing their perceptions of the world. Many continue to find in the horse a source of physical, psychological and spiritual inspiration (Root, 2000).

Further Theoretical Contributions

Theories of attachment (Bowers & McDonald, 2001; Sable, 1995) and neurodevelopment (Perry, 2002; Perry & Hambrick, 2008) have been suggested as applicable to the theoretical framework of EFP. It is an established principle in psychotherapy that the strength of attachment in the helping relationship, characterized by warmth, empathy, trust, acceptance and collaboration is the most powerful predictor of positive client outcome (Lambert & Bergin, 1994; see also Fine, 2000; Schultz, Remick-Barlow & Robbins, 2007; Yorke, 1997). A successful alliance requires a connection and the mobilization of hope, and is facilitated by encouraging a client's sense of well-being and offering insights in a nurturing environment, while encouraging the development of mastery. Horses involved in the therapeutic process can be a useful adjunct in the establishment of the therapeutic relationship and to enhance the development of the alliance.

Horses provide warm and accepting companionship, which provides essential comfort and an unconditional support system to individuals in turmoil (Fine, 2000). Psychotherapy involving horses has the singular advantage of therapeutic touch in a manner not possible with smaller companion animals, or for obvious ethical reasons, with the therapist. Because the horse can be ridden it is possible that neural circuits can be re-wired in an adaptive manner, although care must be exercised with certain populations, for example, clients with a history of trauma or abuse (NARHA, 2008).

Recent research on the therapeutic effects of exercise, particularly with depressed clients, is intriguing in its applicability to the theoretical base of this treatment strategy (Brosse, Sheets, Lett & Blumenthal, 2002; Cripps, 2008; Johnson, 2009; Penedo & Dahn, 2005; Tsang, Chan & Cheung, 2008). The barn as milieu (Hallberg, 2008; Pressly & Heesacker, 2001) has also been put forth as an avenue for investigation, as has risk behavior application (Bailey, 2007; see also Biery, 1985; Mayberry, 1978; Roberts, 1994). All of these areas hold promise as the theoretical underpinnings of equine-facilitated psychotherapy continue to be constructed.

Conclusion

Horsemanship as a therapeutic intervention is discussed in historical context, and the theoretical underpinnings are examined. A framework upon which to base an understanding of this treatment approach is suggested, with the overarching aim of enhancing the credibility of, and thereby to increase scholarly attention to, this unique intervention, ultimately leading to sorely needed research funding and, finally, to third-party reimbursement. Most importantly, this review is intended to encourage further development of the field.

Acknowledgements: Alexa Smith-Osborne was instrumental in providing support for the completion of this project.

References

Acampora, T. (2008). Nothing Short of a Miracle. *Directions*, 4, 8-10.

American Hippotherapy Association (AHA) (n.d.). *History of Hippotherapy and AHA, Inc.* Retrieved May 7, 2008, from http://www.americanhippotherapyassociation.org/ hippotherapy/history-of-hippotherapy/

Bailey, L. (2001). Evolutionary Psychology and Risk Taking. *Medical News Today*, 8, 9-14.

Bain, A. M. (1965). Pony Riding for the Disabled. *Physiotherapy*, 51(8), 263-265.

Barker, S. & Dawson, K. (1998). The Effects of Animal Assisted Therapy on Anxiety Ratings of Hospitalized Psychiatric Patients. *Psychiatric Services*, 49(6), 797-801.

Barker, S. B. & Wolen, A. R. (2008). The Benefits of Human-Companion Animal Interaction: A Review. *Journal of Veterinary Medical Education*, 35(4), 487-495.

Batson, K., McCabe, B., Baun, M., & Wilson, C. (1998). The Effect of a Therapy Dog on Socialization and Physiological Indicators of Stress in Persons Diagnosed with Alzheimer's Disease. In C. Wilson & D. Turner (Eds.), *Companion Animals in Human Health* (pp. 203-215). Thousand Oaks, CA: Sage Publications, Inc.

Beck, A. & Katcher, A. (1996). *Between Pets and People: The Importance of Animal Companionship.* Revised Edition. West Lafayette, IN: Purdue University Press.

Benda, McGibbon & Grant (2003). Improvements in Muscle Symmetry in Children with Cerebral Palsy after Equine-Assisted Therapy (Hippotherapy). *The Journal of Alternative and Complimentary Medicine*, 9(6), 817-825.

Benjamin, J. (2000, Summer). *Introduction to Hippotherapy*. American Hippotherapy Association. Retrieved 9/ 15/2006, from http://www.americanhippotherapyassociation. org/aha_hpot_A-intro.htm

Bertoti, D. B. (1988). Effects of Therapeutic Horseback Riding on Posture in Children with Cerebral Palsy. *Journal of the American Physical Therapy Association*, 68(10), 1505-1512.

Besthorn, F. H. & Saleebey, D. (2003). Nature, Genetics and the Biophilia Connection:Exploring Linkages with Social Work Values and Practice. *Advances in Social Work*, 4(1), 1-18.

Bieber, N. (1998). Horseback Riding for Individuals with Disabilities: A Personal Historical Perspective. In B. T. Engel (Ed.) *Therapeutic Riding I: Strategies for Instruction, Part 1*. Durango, CO: Barbara Engel Therapy Services.

Biery, M. J. (1985). Riding and the Handicapped. *Veterinary Clinics of North America Small Animal Practice*, 15(2), 345-354.

Blake, H. (1975). *Talking with Horses*. Trafalgar Square Publishing: North Pomfret, VT.

Borzo, G. (2002). *Horse Power: When Riding Turns into Treatment*. American Hippotherapy Association. Retrieved September 15, 2008, from http://www.americanhippotherapy association.org/aha_hpot_A-horsepower.htm

Bossard, J. (1944). *The Mental Hygiene of Owning a Dog*. Mental Hygiene, 28, 408-413.

Bracher, M. (2000). Therapeutic Horse Riding: What Has This to do with Occupational Therapists? *British Journal of Occupational Therapy*, 63(6), 277–282

Brosse, A. L., Sheets, E. S., Lett, H. S. & Blumenthal, J. A. (2002). Exercise and the Treatment of Clinical Depression in Adults: Recent Findings and Future Directions. *Sports Medicine*, 32(12), 741-760.

Butt, E. G. (1981/1998). NARHA—Therapeutic Riding in North America...Its First Decade 1970 to 1980. In B. T. Engel (Ed.), *Therapeutic Riding I: Strategies for Instruction, Part 1*. Durango, CO: Barbara Engel Therapy Services.

Chandler, C., K. (2005). *Animal Assisted Therapy in Counseling*. New York, NY: Routledge.

Christian, J. E. (2005). All Creatures Great and Small: Utilizing Equine-Assisted Therapy to Treat Eating Disorders. *Journal of Psychology and Christianity*, 24, 65-67.

Corson, S. & Corson, E. (1979). Pets as Mediators of Therapy. In J. Masserman (Ed.), *Current Psychiatric Therapies*, (pp. 195-205). New York: Grune & Stratton.

Cripps, F. (2008). Exercise your Mind: Physical Activity as a Therapeutic Technique for Depression. *International Journal of Therapy and Rehabilitation*, 15(10), 460-465.

Cushing, J., & Williams, J. (1995). The Wild Mustang Program: A Case Study in Facilitated Inmate Therapy. *Journal of Offender Rehabilitation*, 22(3/4), 95-112.

Database Olympics (2008). *Lis Hartel Olympic Medals and Statistics*. Retrieved February 21, 2009, from http://www. databaseolympics.com/players/playerpage.htm

Debuse D, Chandler C, Gibb C (2005). An Exploration of German and British Physiotherapists' Views on the Effects of Hippotherapy and their Measurement. *Physiotherapy Theory and Practice*, 21 (4): 219-242.

Dell, C. A., Chalmers, D., Dell, D., Sauve, E., & MacKinnon, T. (2008). Horse as Healer: An Examination of Equine Assisted Learning in the Healing of First Nations Youth from Solvent Abuse. *Pimatisiwin: A Journal of Aboriginal and Indigenous Community Health*, 6(1), 81-106.

Delta Society. (1996). *Animal Assisted Therapy: Therapeutic Interventions*. Renton, WA: Delta Society.

DePauw, K. P. (1986). Horseback Riding for Individuals with Disabilities: Programs, Philosophy, and Research. *Adapted Physical Activity Quarterly*, 3, 217-226

Eggiman, J. (2006). Cognitive-Behavioral Therapy: A Case Report—Animal Assisted Therapy. *Topics in Advanced Practice Nursing eJournal*, 6(3). Retrieved October 16, 2006 from: http://www.medscape.com/viewarticle/545439.

Engel, B. T. (1998). Therapeutic Riding: Its Benefits, Professions and Divisions. In B. T. Engel (Ed.), *Therapeutic Riding I: Strategies for Instruction, Part 1*. Durango, CO: Barbara Engel Therapy Services.

Equine Assisted Growth and Learning Association [EAGALA], (2005). Retrieved November 28, 2008, from http://www.eagala.org

Equine Assisted Growth and Learning Association (EAGALA) (n.d). Why Horses? Retrieved March 21, 2007 from http://www.eagala.org

Equine Guided Education Association (n.d.). Retrieved March 26, 2009, from http://www.equine guidededucation.org

Ewing, C. A., MacDonald, P. M., Taylor, M., & Bowers, M. J. (2007). Equine-Facilitated Learning for Youths with Severe Emotional Disorders: A Quantitative and Qualitative Study. *Child Youth Care Forum*, 36, 59-72.

Fine, A., H. (2000). Animals and Therapists: Incorporating Animals in Outpatient Psychotherapy. In A. Fine (Ed.), *Handbook on Animal-Assisted Therapy: Theoretical Foundations and Guidelines for Practice*, (pp. 179-193). San Diego, CA: Academic Press.

Fitzpatrick, J. C. & Tebay, J. M. (1998). Hippotherapy and Therapeutic Riding: An International Review. In C. Wilson & D. Turner, (Eds.), *Companion Animals in Human Health*, (pp. 41-58). London: Sage Publications, Inc.

Folse, E., Minder, C., Aycock, M., & Santana, R. (1994). Animal Assisted Therapy and Depression in Adult College Students. *Anthrozoös*, 7(3), 188-194.

Fosdick, K. L. (2003). *History of Therapeutic Riding*. Retrieved November 11, 2008, from http://www.thesaddlelightcenter.com/therapriding.htm

Fredrickson, M. (1992). Terms and Definitions in the Field of Human-Animal Interactions. In B. Engel (Ed.), *Therapeutic Riding Programs Instruction and Rehabilitation: A Handbook for Instructors and Therapists* (pp. 30-32). Durango, CO: Barbara Engel Therapy Services.

Fredrickson, M. (2008). *Fear and Grace*. Saint Paul, MN: Minnesota Linking Youth, Nature, and Critters, Inc.

Friedmann, E., Thomas, S. A. & Eddy, T. J. (2000). Companion Animals and Human Health: Physical and Cardiovascular Influences. In A. L. Podberscek, E. S. Paul & J. A. Serpell, (Eds.), *Companion Animals and us: Exploring the Relationships between People & Pets* (pp. 125-142). New York: Cambridge University Press.

Frewin, K. & Gardiner, B. (2005). New Age or Old Sage? A Review of Equine Assisted Psychotherapy. *The Australian Journal of Counselling Psychology*, 6, 13-17.

Garrity, T., F. & Stallones, L. (1998). Effect of Pet Contact on Human Well-Being. In Wilson & Turner (Eds.), *Companion Animals in Human Health*, (pp. 3-22). Thousand Oaks, CA: Sage Publication, Inc.

Gatty, C. M. (n.d.). *Psychosocial Impact of Therapeutic Riding: A Pilot Study*. Retrieved December 6, 2008, from http://www.narha.org/PDFfiles/Psychosocial_Impact.pdf

Glasow, B. L. (2007). *Semantics—To be Exuberant or to be Correct*. American Hippotherapy Association. Retrieved March 3, 2009, from http://www.americanhippotherapy association.org/aha_hpot_a_sem.htm

Granger, B. & Granger, G. V. (2004). *Evaluating the Effectiveness of Animal-Assisted Therapy Approaches in an Alternative High School for Expelled Youth: A Qualitative/Quantitative Analysis*. Conference: Can Animals Help Humans Heal: Animal-Assisted Interventions in Adolescent Mental Health.

Grier, K. C. (1999). Childhood Socialization and Companion Animals: United States, 1820-1870. *Society and Animals*,7, 95-120.

Griffith, J. C. (1992, Spring). Chronicle of Therapeutic Horseback Riding in the United States, Resources and References. *Clinical Kinesiology*, 2-7.

Hallberg, L. (2008). *Walking the Way of the Horse: Exploring the Power of the Horse-Human Relationship*. iUniverse, Inc.: New York, NY.

Hansen, K., Messinger, C., Baun, M., & Megel, M. (1999). Companion Animals Alleviating Distress in Children. *Anthrozoös*, 12,(3), 142-148.

Harpøth, U. (1970). Horseback Riding for Handicapped Children. *Physical Therapy*, 50(2), 235-236.

Haskin, Erdman, Bream & MacAvoy (1974). Therapeutic Horseback Riding for the Handicapped. *Archives of Physical Medicine and Rehabilitation*. 55(10), 473-474.

Hayden, A. J. (2005). *An Exploration of the Experiences of Adolescents who Participated in Equine Facilitated Psychotherapy: A Resiliency Perspective*. Unpublished doctoral dissertation, Alliant International University, San Diego, CA. (UMI No. 3156902).

Haynes, M. (August, 1991). Pet Therapy: Program Lifts Spirits, Reduces Violence in Institution's Mental Health Unit. *Corrections Today*, 120-122.

Henriksen, J. D. (1971). Horseback Riding for the Handicapped. *Archives of Physical Medicine and Rehabilitation*, 52(6), 282-283.

Hines, L. (1983). Pets in Prison: A New Partnership. *California Veterinarian*, May, 7-17.

Hines, L. (2003). Historical Perspectives on the Human-Animal Bond. *American Behavioral Scientist*, 47, 7-15.

Holcomb, R., Jendro, C., Weber, B., & Nahan, U. (1997). Use of an Aviary to Relieve Depression in Elderly Males. *Anthrozoös*, 10(1), 32-36.

Irwin, C. & Weber, B. (2005). *Dancing with your Dark Horse: How Horse Sense Helps us Find Balance, Strength and Wisdom*. New York: Marlow & Company.

Johnson, C. C. (2009). The Benefits of Physical Activity for Youth with Developmental Disabilities: A Systematic Review. *American Journal of Health Promotions*, 23(3), 157-167.

Karol, J. (2007). Applying a Traditional Individual Psychotherapy Model to Equine-Facilitated Psychotherapy (EFP): Theory and Method. *Clinical Child Psychology and Psychiatry*, 12, 77-90.

Katcher, A. (2002). Animals in Therapeutic Education: Guides into the Liminal State. In P. Kahn & S. Kellert (Eds.), *Children and Nature: Psychological, Sociocultural and Evolutionary Investigations*. Cambridge, MA: The MIT Press.

Katcher, A, Beck, M., & Levine, D. (1989). Evaluation of a Pet Program in Prison: The PAL Project at Lorton. *Anthrozoös*, 2(3), 175-180.

Katcher, A. & Wilkins, G. G. (1998). Animal Assisted Therapy in the Treatment of Disruptive Behavior Disorders in Children. In A. Lundberg (Ed.), *The Environment and Mental Health* (pp. 193-202). Mahwah, NJ: Lawrence Erlbaum Associates.

Kellert, S. R. (1996). *Kinship to Mastery: Biophilia in Human Evolution and Development*. Washington, DC: Island Press.

Kellert, S. R. & Wilson, E. O. (1993). *The Biophilia Hypothesis*. Washington, DC: Island Press.

Kersten, G. W. (1997). Equine Assisted Psychotherapy. In R. Harwell, T. Comstadt, & N. Roberts (Eds.), *Deeply Rooted, Branching out*, 1972-1997: Annual AEE International Conference Proceedings. Boulder, CO: Association for Experiential Education.

Kohanov, L. (2001).*The Tao of Equus*. Novato, CA: New World Library.

Kroeger, A. (1992). Using Vaulting Lessons as Remedial Education. Pp. 609-617 in B. Engel (Ed.), *Therapeutic Riding Programs: Instruction and Rehabilitation*. Durango, CO: Barbara Engel Therapy Services.

Kruger, K. A., Trachtenberg, S. W., & Serpell, J. A. (2004). *Can Animals Help Humans Heal?* Animal-Assisted Interventions in Adolescent Mental Health. Report from the Center for the Interaction of Animals and Society, University of Pennsylvania School of Veterinary Medicine. Retrieved April 28, 2007, from http://www2.vet. upenn.edu/research/centers/ cias/pdf/CIAS_AAI_ white_paper.pdf

Lambert, M. J. & Bergin, A. E. (1994). The Effectiveness of Psychotherapy. In A. Bergin & S. Garfield (Eds.), *Handbook of Psychotherapy and Behavior Change* (pp. 143-190). NY, NY: John Wiley & Sons.

Levinson, B. M. (1962). The Dog as a Co-Therapist. *Mental Hygiene*, 46, 59-65.

Levinson, B. M. (1965). Pet Psychotherapy: Use of Household Pets in the Treatment of Behavior Disorder in Childhood. *Psychological Reports*, 17, 695-698.

Levinson, B. M. (1969a). Household Pets in Residential Schools: Their Therapeutic Potential. *Mental Hygiene*, 53(3), 411-415.

Levinson, B. M. (1969b). Pets and Old Age. *Mental Hygiene*, 53(3), 364-368.

Levinson, B. M. (1971). Household Pets in Training Schools Serving Delinquent Children. *Psychological Reports*, 28, 475-481.

Louv, R. (2006). *Last Child in the Woods: Saving our Children from Nature-deficit Disorder*. Chapel Hill, NC: Algonquin Books of Chapel Hill.

MacKinnon, J. R., Noh, S., Laliberte, D., Lariviere, J. & Allan, D. E. (1995). Therapeutic Horseback Riding: A Review of the Literature. *Physical and Occupational Therapy in Pediatrics*, 15(1), 1-16.

Mayberry, R. P. (1978). The Mystique of the Horse is Strong Medicine: Riding as Therapeutic Recreation. *Rehabilitation Literature*, 39(6-7), 192-196.

McDaniel, I. (1998, Winter). What Exactly is "Equine Facilitated Mental Health & Equine Experiential Learning?" *Strides*, 30-31.

McPhail, K. L. (2006, Summer). Occupational Therapy at the Barn. Evidence Through Measurement: Outcome Measures for Hippotherapy Research. *Hippotherapy*.

McVarish, C. (1995). The Effects of Pet Facilitated Therapy on Depressed Institutionalized Inpatients. *Dissertation Abstracts International*, 55(7-B), 3019.

Meregillano, G. (2004). *Hippotherapy*. Physical Medicine and Rehabilitation Clinics of North America, 15(4), 843-854.

Moneymaker, J. M., & Strimple, E. O. (1991). Animals and Inmates: A Caring Companionship Behind Bars. *Journal of Offender Rehabilitation*, 16(3-4), 133-152.

Moreau, L. M. & McDaniel, I. (2000). *Equine Facilitated Mental Health: A Field Guide for Practice*. 2nd ed.

Morris, D. (1988). *Horsewatching: Why Does a Horse Whinny and Everything Else You Ever Wanted to Know*. New York: Crown Publishers, Inc.

Netting, F. E., Wilson, C. C. & New, J. C. (1987, January-February). The Human-Animal Bond: Implications for Practice. *Social Work*, 60-64.

North American Riding for the Handicapped Association (NARHA). (n. d.). Retrieved April 10, 2006, from http://www.narha.org/

NARHA Equine-facilitated Mental Health Association (EFMHA). (2008). *EFP Precautions*, 1-2. Retrieved May 5, 2009, from http://www.narha.org/SecEFMHA/Precautions.asp

NARHA EFMHA. (2008). *EFP Contraindications*, 1. Retrieved May 5, 2009, from http://www.narha.org/SecEFMHA/Contraindications.asp

Ormerod, E. (1996). Pet Programmes in Prisons. *The SCAS Journal*, Winter, 8(4), 1-3.

Penedo, F. J. & Dahn, J. R. (2005). Exercise and Well-Being: A Review of Mental and Physical Health Benefits Associated with Physical Activity. *Behavioural Medicine* 18(2), 189-193.

Perry, B. D. (2002). Childhood Experience and the Expression of Genetic Potential: What Childhood Neglect Tells us about Nature and Nurture. *Brain and Mind*, 3, 79-100.

Perry, B. D. & Hambrick, E. P. (2008). The Neurosequential Model of Therapeutics. *Reclaiming Children and Youth*, 17, 38-43.

Politiken.DK. (2009). *Obituary: DK Equestrian Legend Lis Hartel*. 13 February 2009. Retrieved February 21, 2009, from http://www.politiken.dk/newsinenglish/article649135.ece? service

Pressly, P. K. & Heesacker, M. (2001). The Physical Environment and Counseling: A Review of Theory and Research. *Journal of Counseling & Development*, 79, 148-160.

Professional Association of Therapeutic Horsemanship International (PATH Intl.). (2011, June 20). NARHA is now Path Intl.. Retrieved 6/20/2011, from www.pathintl.org/

Rashid, M. (2000). *Horses Never Lie*. Pagosa Springs, CO: Johnson Books.

Rector, B. (n.d.). *Adventures in Awareness*. Retrieved January 28, 2012, from http://www.adventuresinawareness.net/ sessions.htm

Reichert, E. (1998). Individual Counseling for Sexually Abused Children: A Role for Animals and Storytelling. *Child and Adolescent Social Work Journal*, 15, 177-185.

Roberts, F., Bradberry, J., & Williams, C. (2004). Equine-Facilitated Psychotherapy Benefits Students and Children. *Holistic Nursing Practice*, 18, 32-35.

Roberts, P. (1994, November/December). Risk. *Psychology Today*.

Robin, M. & ten Bensel, R. (1985). Pets and the Socialization of Children. *Marriage and Family Review*, 8, 63-78.

Root, A. (2000). *Equine-Assisted Therapy: A Healing Arena for Myth and Method*. Unpublished Doctoral Dissertation. Pacifica Graduate Institute.

Russell-Martin, L. A. (2006). *Equine Facilitated Couples Therapy and Solution Focused Couples Therapy: A Comparison Study*. Unpublished doctoral dissertation, Northcentral University, Prescott, Arizona.

Scanlan, L. (1998). *Wild About Horses: Our Timeless Passion for the Horse*. New York, NY: HarperCollins.

Schultz, P. N., Remick-Barlow, G. A., & Robbins, L. (2007). Equine-Assisted Psychotherapy: A Mental Health Promotion/Intervention Modality for Children who have Experienced Intra-Family Violence. *Health and Social Care in the Community*, 15, 265-271.

Scott, N. (2005). *Special Needs, Special Horses: A Guide to the Benefits of Therapeutic Riding*. Denton, TX: University of North Texas Press.

Serpell, J. A. (2000). Animal Companions and Human Well-Being: An Historical Exploration of the Value of Human-Animal Relationships. In A. Fine (Ed.), *Animal Assisted Therapy*, (pp. 3-19). San Diego, CA: Academic Press.

Silkwood-Sherer, D. & Warmbier, H. (2007). Effects of Hippotherapy on Postural Stability in Persons with Multiple Sclerosis: A Pilot Study. *Journal of Neurologic Physical Therapy*, 31(2), 77-84.

Sloan, A. W. (1987). Thomas Sydenham, 1624-1689. *South African Medical Journal*, 72, 275-278.

Snider, L., Korner-Bitensky, N., Kammann, C., Warner, S. & Saleh, M. (2007). Horseback Riding as Therapy for Children with Cerebral Palsy: Is There Evidence of its Effectiveness? *Physical & Occupational Therapy in Pediatrics*, 27, 5-23.

Sterba, J. A., Rogers, B. T., France, A. P. & Vokes, D. A. (2002). Horseback Riding in Children with Cerebral Palsy: Effect on Gross Motor Function. *Developmental Medical Child Neurology*, 44, 301-308.

Taylor, S. M. (2001). *Equine Facilitated Psychotherapy: An Emerging Field*. Unpublished master's thesis, Saint Michael's College, Colchester, Vermont.

Thomas, L. (2001). More than Magic. *Equine Assisted Growth and Learning News*, 5-6.

Thomas, L. (2002). Horse-Play can be Therapeutic: Equine-Assisted Psychotherapy. *Woodbury Reports*, 100, 69-73.

Tsang, H. W., Chan, E. P. & Cheung, W. M. (2008). Effects of Mindful and Non-Mindful Exercises on People with Depression: A Systematic Review. *British Journal of Clinical Psychology*, 47, 303-322.

Ulrich, R. S. (1993). Biophilia, Biophobia, and Natural Landscapes. In S. Kellert & E. Wilson (Eds.), *The Biophilia Hypothesis* (pp 73-137). Washington, D. C.: Island Press.

Vidrine, M., Owen-Smith, P., & Faulkner, P. (2002). Equine-Facilitated Group Psychotherapy: Applications for Therapeutic Vaulting. *Issues in Mental Health Nursing*, 23, 587-603.

West, J. (1970). Horse Riding for Mentally Handicapped Children. *Nursing Mirror*, 18, 13-14.

White, R. & Heerwagen, J. (1998). Nature and Mental Health: Biophilia and Biophobia. In A. Lundberg (Ed.), *The Environment and Mental Health* (pp. 175-192). Mahwah, NJ: Lawrence Erlbaum Associates.

Wilson, E. O. (2007). Biophilia and the Conservation Ethic. In D. J. Penn & I. Mysterud (Eds), *Evolutionary Perspectives on Environmental Problems* (pp. 249-257). New Brunswick, NJ: Transaction Publishers.

Wingate, L. (1982). Feasibility of Horseback Riding as a Therapeutic and Integrative Program for Handicapped Children. *Physical Therapy*, 62(2), 184-186.

Yorke, J. (1997). *The Therapeutic Value of the Equine-Human Relationship in Recovery from Trauma: A Qualitative Analysis*. Unpublished Master's thesis. Wilfred Laurier University, Ottawa, ONT, CAN.

Yorke, J. (2008). The Significance of Human-Animal Relationships as Modulators of Trauma Effects in Children: A Developmental Neurobiological Perspective. *Early Child Development and Care*, 1-12.

Yorke, J., Adams, C., & Coady, N. (2008). Therapeutic Value of Equine-Human Bonding in Recovery from Trauma. *Anthrozoös*, 21(1), 17-30.

Article Two

The EAGALA Model of Equine Assisted Psychotherapy and Learning

Lynn Thomas, LCSW, Co-founder & Executive Director, EAGALA
Correspondence to:
Lynn Thomas, Executive Director EAGALA
PO Box 993
Santaquin, UT 84655, USA
Phone: +1.801.754.0400
Email: LThomas@eagala.org
Website: www.eagala.org

Abstract

The Equine Assisted Growth and Learning Association (EAGALA), established twelve years ago has a unique approach to equine assisted psychotherapy and learning which is based firmly on sound professional principles in horsemanship, psychotherapy and pedagogy. This paper outlines the development and basis of the EAGALA model, and also provides practical examples of its application in practice.

Introduction

The Equine Assisted Growth and Learning Association (EAGALA) was founded in Santaquin, Utah, USA as a nonprofit organization to address the need for resources, education and professionalism in Equine Assisted Psychotherapy (EAP) and Equine Assisted Learning (EAL). Since its inception in 1999, EAGALA has grown to over 3,500 members in 40 countries. The growth in EAGALA is unique in the nonprofit industry and is indicative of the growing interest in horses as a modality for mental health treatment and personal and professional development.

What is EAGALA?

EAGALA is a membership association which establishes standards of practice, provides annual conferences, training and certification in EAP/EAL, publishes EAGALA In Practice magazine, has a research committee which encourages and collects research, and offers networking and educational resources.

EAGALA supports its members worldwide through Regional Coordinators who oversee five established regions – Africa, Europe/Middle East (EME), Latino America, Pacific (covering Asia, Australia and New Zealand) and North America. Within these regions have been established 73 local networking groups, each with a coordinator, who holds meetings for local EAGALA members for practice, peer support and collaborative efforts.

EAGALA's management team includes Executive Director and Co-founder, Lynn Thomas, Licensed Clinical Social Worker, and Program Director, Mickey DiGiacomo. A volunteer board of directors governs the nonprofit organization ensuring longevity of the association, efficient use of resources, focus on EAGALA's mission and objectives, and high standards of business practice.

What is the EAGALA Model?

As the interest in EAP/EAL grows, different ideas, organizations and approaches also surface as in any industry. EAGALA's training and certification program is based on the EAGALA Model which provides a standard and structure for providing EAP/EAL sessions. Practicing within a model establishes a foundation of key values and beliefs, and provides a basis of good practice and professionalism. Practicing within a model also supports research efforts by providing a standard and ability for replication. The EAGALA Model provides a framework of practice, but within that framework there are infinite opportunities for creativity and adaptability to various therapeutic styles.

The EAGALA Model has 4 key tenets as its standard:

- **The Team Approach** – An Equine Specialist, a Mental Health professional, and horses work together with clients in all EAGALA sessions. The Equine Specialist and Mental Health professional must each meet minimum standards to serve in these roles (see Certification information below).
- **Focus on the ground** – No horseback riding is involved. Instead, effective and deliberate techniques are utilized where the horses are metaphors in specific ground-based experiences.
- **Solution-Oriented** – The belief that our clients have the best solutions for themselves forms a foundation for the EAGALA approach. Rather than instructing or directing solutions, we allow our clients to experiment, problem-solve, take risks, employ creativity, and find their own solutions.
- **Code of Ethics** – EAGALA has high standards of practice and ethics and an ethics committee and protocol for upholding these standards – ensuring best practices and the highest level of care. (see http://eagala.org/Ethics)

Roles of the EAGALA Team

- **The Horse**: Horses have many characteristics which lend them to being effective agents of change, including honesty, awareness, and ability with nonverbal communication. The role of the horses in an EAGALA session is to be themselves – which is one of the reasons EAGALA's model involves horses at liberty with no riders.
- **The Equine Specialist (ES)**: The ES chooses the horses to be used in sessions, works with the MH to develop the session structure, keeps an equine log to document horse behaviors in sessions, stays aware of safety and welfare of clients, horses, and team, and makes observations of horses which can bring in potential metaphors.
- **The Mental Health Professional (MH)**: The MH is responsible for treatment planning, documentation of clients, and ensuring ethical practice. The MH builds on the ES's horse observations, bringing in the metaphoric and therapeutic/learning relevance of the session.

FACILITATION

While therapeutic riding has the goal of teaching riding with its accompanying therapeutic benefits, and hippotherapy has the goal of physical, occupational or speech and language therapy, EAGALA's training program teaches Equine Specialists and Mental Health professionals how to work together to structure and facilitate effective sessions to meet mental health and learning goals. Whether clients learn about horses is not the focus, although it may be a by-product.

Sessions are structured to provide a parallel to the clients' lives. The horses become metaphors in what the clients are working on related to their therapeutic and learning goals. Theoretical foundations to facilitating EAGALA Model sessions include experiential principles (see www.aee.org/about/whatIsEE), a client-centered approach (Rogers 1951) where the facilitator listens to and relies on the clients and horses to direct the sessions, using projection and metaphor. The facilitator avoids placing his/her interpretations and agendas on the clients and the process (the Clean Language model by David Grove is a great example of this process.). Rather, the focus for facilitators is to reflect observations and ask questions. They do this from a place of genuine curiosity which invites reflective thought and does not direct clients to the facilitator's biased solutions (Sullivan & Rees, 2008).

To assist in this process, EAGALA created a framework called SPUD'S (trademarked by EAGALA). It assists facilitators on where to focus observations to be reflected to the clients and provide opportunities for insight and metaphor. The observations must be objective, non-interpretive, behavioral (using Clean Language), and primarily focused on the horses.

S – Shifts: Watch for any shifts that occur in behaviors, including actual physical placement, of both horses and humans. For instance, the horses were standing, now they are running.

P – Patterns: A pattern is when a behavior occurs three or more times, and these patterns tend to parallel other patterns in the clients' lives.

U – Unique: When observing, if the horses do something out of the ordinary, like kick up, or something dramatic or unusual, or if the humans do something that is quite unique, note that moment - what the unique behavior was, and what was happening in the entire arena with humans and horses in that moment.

D – Discrepancy: Listen to the verbal communication of the clients and compare it to the non-verbal. For instance, the client says she "doesn't want to do this," but at the same time she is walking towards the horses. The primary reason for paying attention to these discrepancies is because it indicates an opportunity for change - a motivation – and a potential metaphor.

'S (apostrophe S) - Self-awareness or "My Stuff": While observing the experience of the clients and horses, facilitators must constantly observe counter transference, i.e. observing ourselves. How does 'S influence our observations and processing, i.e. agenda-filled observations and questions? Through the years of training facilitators doing this work, we have found that the 'S has more impact on the outcomes of sessions with clients than probably any other factor.

SESSION EXAMPLE

The EAGALA Model involves structuring sessions to reflect real-life issues which provide the clients an opportunity to experience and explore their current behaviors and beliefs, and discover those which may be more effective for healthy living.

The following is an example of an EAGALA Model session with military service members designed to address the psychological challenges of redeployments:

Participants are asked to build two spaces utilizing various props in an arena setting; one space represents their time being deployed, the second home or life in the garrison. What they build and the meaning behind it is then discussed. For example, the props may represent physical and emotional reminders of their deployment and can be labeled accordingly. Also discussed is how the horses in the arena responded to their building each space – and the meaning that may have to their process.

The service members are then asked to move a group of horses from the home space to the deployment space and back – representing their deployment transitions. This in itself presents them with a number of difficulties. Horses have different personalities, may not want to get separated from each other or might not want to be brought to a certain area. How the soldiers go about this activity – which horses they pick and how they communicate with each other and the horses – each aspect of this process and the horses' responses to their choices is then discussed and related to parallels in their transition experience.

The service members will determine what metaphor the horses become for them: the horses may become a representation of themselves or family members, or they may represent feelings, expectations, strengths, dreams or hopes. By creating an experience like this, the service members have the opportunity to project on to the horses and experience real-life concerns, patterns, and beliefs. By addressing the metaphors, the behaviors of the horses and experienced difficulties and successes, participants have the opportunity to draw parallels to how they react in similar situations with similar difficulties. They not only talk about challenges and solutions, they experience it in a hands-on way providing opportunity for powerful and meaningful learning.

Certification Program

EAGALA Certification can be obtained by successfully completing two three-day trainings in Fundamentals of the EAGALA Model Practice (Part 1 and Part 2) and submitting a Professional Development Portfolio for review. EAGALA Advanced Certification necessitates additional training and requirements, including participation in the mentoring program. Certification renewal is required every two years and involves continuing education.

EAGALA Certified members commit to standards of practice and ethics, which include a minimum requirement of experience and education for the facilitating team. The Equine Specialist must have at least 6,000 hours hands-on work with horses, along with over 100 hours of continuing education. The Mental Health Professional must follow regional laws for mental health practice, have education in the mental health scope, and be licensed, registered or certified (depending on country) by a governing body.

Session Logistics

The EAGALA Model can be applied to any treatment or learning focus. It has been used with clients with goals and issues ranging from depression, eating disorders, addictions, schizophrenia, autism, PTSD and family relationships, to educational goals such as life skills, organizational development and leadership. Just as any therapeutic model, program length and frequency varies, however weekly hour or two hour long sessions for 8 – 12 weeks is a common average.

Because the EAGALA Model does not involve riding, it opens the opportunity for more types of horses to be involved in sessions. Retired competitive horses that may no longer be ridden can have a role and purpose through helping people in this method, providing healing opportunities for horses and humans alike. Likewise, programs do not need as much overhead as riding equipment is unnecessary and number of horses can be fewer since it is not necessary to have a horse for every client. EAGALA has established a partnership with the Thoroughbred Retirement Foundation to support finding homes for retired racehorses in EAGALA programs.

Research

Research is still minimal in the EAP/EAL industry and is an area of needed focus. Researchers can go to the EAGALA website and find a bibliography as well as some links to full papers (http://eagala.org/Research). Likewise, with over 400 programs practicing the EAGALA Model worldwide with various types of clients, EAGALA can help researchers connect with programs to conduct research (the Find a Program feature on the EAGALA website enables searching by location or key words). The EAGALA Research Committee can provide resources and answer questions, and they have also developed a single subject research design EAGALA members and researchers may utilize. The website also has a bibliography of measurement tools commonly utilized in outcome research.

Conclusion

Equine Assisted Psychotherapy and Equine Assisted Learning are growing worldwide and provide an interesting and rewarding new niche in both the horse and mental health industries. As with any profession, different approaches and techniques will surface as the modalities continue to evolve. EAGALA provides a clear and replicable model and standard of practice to ensure credibility, professionalism and accountability in this new and growing discipline. Go to www.eagala.org for more information.

References

Association for Experiential Education: http://www.aee.org/about/whatIsEE

EAGALA. *Fundamentals of EAGALA Model Practice Training Manual*. Sixth Edition. Utah: EAGALA; 2009.

Lawley, J & Tompkins, P. (2000). *Metaphors in Mind: Transformation through Symbolic Modelling*. London: Developing Company Press.

Rogers, C. (1951). *Client-centered Therapy: Its Current Practice, Implications and Theory*. London: Constable.

Sullivan, W & Rees, J. (2008). *Clean Language: Revealing Metaphors and Opening Minds*. Carmarthen: Crown House.

ARTICLE THREE

The Frossarbo Therapeutic Model (FTM) – An Experiment within the Institutional Care System of Juvenile Delinquents in Sweden

Sven Forsling, Psychologist, Director, Frossarbo Stables.
Former: Principal Psychologist, Section 12 homes, Metropolitan Authorities of Stockholm, Sweden; Member, Scientific Governing Body National Board of Institutional Care (SiS), Sweden

Correspondence to:
Sven Forsling
Address: Stalbo 642, 740 45 Tärnsjö, Swden
Phone: 0046-0292-41042
Email: sven.forsling@telia.com

ABSTRACT

In the 1980s a Swedish study, "444 Stockholm Youth in Crisis," identified the special institutional needs and challenges of young females aged 16 – 21 years. It pointed to the need for change in the prevailing high cost, staff intensive closed unit system with its poor therapeutic outcomes. In response the Frossarbo Therapeutic Model (FTM), based on a psychodynamic therapeutic approach and Milieu Therapy, was carried out and evaluated between 1987 and 1997. This model went against existing practice and focused on the strength of the juveniles instead of on their distress diagnoses. Over the 10 year period 47 girls (6 at a time) came to Frossarbo Stables for an alternative therapy which involved learning about, caring for and training trotting horses in preparation for racing. FTM was evaluated against standard variables established by The General Board of Compulsory Treatment, Stockholm Metropolitan Council. It received top ratings in all variables.

KEYWORDS

Compulsory Treatment; Juvenile Delinquents; Delinquent Girls; Institutional Care; Psychodynamic Therapy; Milieu Therapy; 24 Hour Therapy; Frossarbo Therapeutic Model; FTM; Equine Assisted Therapy; EAT; Animal Assisted Therapy; AAT.

Background

In 1983, responsibility for institution-based custodial care of young persons with serious social problems was transferred from the State to County and District councils. The care, given generally without the co-operation of the individual involved, was regulated by the Care of Young Persons Act. The youngsters in these institutions (often referred to as Section 12 homes) displayed a broad range of deep psychosocial problems with criminality and drug abuse being dominant. Every year, about 100 to 120 youths were sent to these specially approved institutions. Their ages ranged from 12 to 21 years, but the majority were between 15 and 17 years of age. There was one girl to every four or five boys. The responsibility for custodial care of juveniles returned to the State in 1994. It is now run by the National Board of Institutional Care (SiS).

This experiment in "alternative" institutional care was carried out between the years 1987 to 1997. It was a direct result of a follow up study, "444 Stockholm youth in Crisis" (Forsling, 1987), that among other things pointed to the need for:

- a "girls only" institution in the age span 16 – 21 years. The study had shown (later confirmed by Andersson, 1998 and Runquist, 1999) that delinquent girls placed together with boys, had limited possibilities of tackling their own problems. The girls tended to listen more to the boys, giving them priority. They also ran a high risk of sexual abuse.
- new ways to tackle the challenge of institution care. The prevailing view was for more closed units which involved more staff, higher costs and little therapeutic gain.

At this time I was Principal Psychologist and Director of Research for Custodial Care within Stockholm and Stockholm County and was worried about the development of the Section 12 homes. I felt they had ossified in a straitjacket of rules and regulations. One could spend hours at conferences talking about how wretched our youngsters had become and how this wretchedness should be dealt with yet, despite staff increases, the time available to spend with the youngsters was decreasing and relationships worsened rather than improved.

Compulsory institutional treatment is based on the apparently contradictory principles of providing a situation that both:

- fosters psychological growth, and
- applies institutional retention, often by force.

How under these circumstances do you build trust?

The experiment at Frossarbo Stables between 1987 – 1997 was designed to see what would happen if we broke the vicious circle of more and more restrictions; if we built on the juveniles' strength instead of focusing on their shortcomings; if we shared interest in a close working situation with them where their capacity as well as ours was needed to carry through the work. Could we call for trust instead of distrust? Could we leave the doors unlocked and tear down the barriers if staff became the only walls and locks? Would it be possible to run such an institution?

The aim was to assess the possibility of achieving security and therapeutic progress by an alternative method – namely, by providing choice within the compulsory system, using fewer personnel but a close working relationship with the juveniles to see if this would increase trust and therapeutic progress.

THE FROSSARBO THERAPEUTIC MODEL (FTM)

The therapeutic model tested at Frossarbo Stables centred its treatment on girls and horses. It trained 16-21 year old delinquent girls, six at a time, to care for horses and taught them the many preparation stages for horses being trained to race – all designed for the psychological treatment of the girls themselves.

The underlying FTM theoretical approach was based on a psychodynamic approach in therapy (Winnicott, 1971) and Milieu Therapy or 24 hour Therapy (Jones, 1968). An important FTM guideline was that everything should be "for real" – FTM juveniles would work with, learn about and train real trotting racehorses.

Juveniles with little motivation, low self-esteem and few words, present a real challenge to psychodynamic therapy. FTM aimed to break down the walls of traditional therapy, so often locked into a room of its own making, confined by its own methods, constrained by a specific time and a restrictive place and with a client who was also locked into some specific ways of reacting. Its objective was to find a format where even the destructive and seriously disconnected individual (those who made a mess of everything and those who rejected everything) would have a chance. The potential number of possible clients had to be very clearly restricted.

In essence, shared work and stable life was FTM's therapy. Therapists put on overalls and got to work with youngsters and horses. The everyday moments of challenge, conflict and joy were therapy, providing opportunities to learn and build relationships between the therapists, the youngsters and the horses. Assessments and evaluations developed naturally through genuine trust.

The horse activity basis of FTM was consistent with Milieu Therapy – if you are enthusiastic and like something then there is a good possibility that those you work with will start to like that too. You share something of yourself, something that is for real, important in any kind of institutional treatment. The environment at an institution can easily become artificial with too many artificial rules.

I chose to build FTM around horses. I made this choice, not because of the recognized horse interest among girls, but because this was a strong interest of my own. My knowledge and interest, if passed on to the girls, would carry with it the possibility to learn a real trade and to provide a real job. I had a relevant background from working in a race stable in my youth, I was a licensed blacksmith and I had a license to race and train harness horses. FTM's approach falls well within the area of Equestrian Assisted Therapy or Animal Assisted Therapy although in the 1980's that name was unknown in Nordic countries.

FTM offered a Horse Management Course comprising a qualified theoretical and practical program in taking care of horses – special horses for racing. It was divided into 4 stages. The first 2 stages, each lasting 2 months, formed the groundwork for the course. The next 2 stages, each spanning 4 months, covered the more advanced work. After that, there were in-depth specialised courses such as horseshoeing and horse psychology. Each in-depth course lasted 4 months.

Every stage and course ended with a written exam. The student received a certificate at the end of every successful examination. Fully completed courses qualified for a diploma at college level and opened up job possibilities at racing stables.

FTM was based on choice. Any girl aged 16 – 21 years sentenced to compulsory treatment had the option to come to Frossarbo Stables if a place was available. Choice was important. It symbolised a crucial step in any treatment, that of daring to say a first vulnerable "yes", especially important in compulsory treatment where the juveniles usually start with a strong "no". This freedom of choice followed the girls throughout their stay. After each course they had to choose again and their "yes" was just for the next course. Another "yes" would have to follow. The shorter length of the first 2 courses was in recognition that coming to a new and different environment is often scary and lonely. To have to commit to a stay of a year, for example, could easily be regarded as forever by many youngsters and force them to give up even before trying.

Each girl took care of one horse, a training horse which was her responsibility throughout her education at Frossarbo. She chose her horse from among those available. A training horse was worked with everyday in preparation for racing. Frossarbo Stables had two legendary trainers as mentors, Sören Nordin and Delvin Miller. This emphasized that the racing was for "real" and also gave extra status to the education. Race winnings went to a fund from which the girls could apply for money when they left Frossarbo.

Keywords in FTM were "for real", "take your time", "see what your choices are and choose", "we all have to help and we are all needed if our horses will have the best", "you will always be allowed to take your time even when getting to know your horse", and "no one will push you."

A full account of the FTM is available in *The Girl and the Horse* (Forsling, 2003).

Evaluation Method and Results

Objective: The aim was to compare FTM's performance with the norms of traditional institutions within the compulsory system – norms established by that system. Most expected Frossarbo Stables to fail on all comparative norms.

Time Period: The 10 years of FTM operation, 1987-1997. FTM challenged the whole system. The study was originally planned to run for 3 years in order to give the experiment time for a real evaluation, however, as it turned out, results are available for all years that FTM operated. During that time the compulsory treatment of youth was first the responsibility of the Stockholm County and then the State Government, (SiS).

Participants: The participant group was composed of girls aged 16 to 21 years sentenced to institutional care who chose to come to Frossarbo Stables.

Evaluation Criteria: Eight objective criteria (and norms against which each was to be assessed) were decided by the Director General of Compulsory Treatment, Stockholm Metropolitan Council through his "Section 12 Secretariat".

In addition, I asked the Secretariat to add:

- a "best" for each norm/variable (to illustrate fluctuations that might be of value when the final decision on the future of Frossarbo Stables was taken).
- four extra criteria specifically related to assessing the FTM's approach of "pushing boundaries" and exploring what could be done irrespective of the juveniles´ diagnosis. I wanted these "boundary pushing" criteria although they had no counterparts in the other Section 12 Homes.

We also agreed that some kind of therapeutic follow up should be done directly after the experiment although the time span would be short.

Section 12 Standards Assessment Criteria

1. Placement Inquiries

- **Norm**: Frossarbo Stables was to be an open-treatment alternative in compulsory care, with places for 6 pupils from/on their way to a closed unit. The average yearly intake was to be 4.5 placements; more would be rated "best".

- **Comment**: This was a crucial criteria. In an organisation, if one component does not deliver, some other part has to take its work. If Frossarbo Stables did not take its share another institution had to carry an extra burden.
- **Result**: "best". After the first 6 months and up to the end of the tenth year, Frossarbo Stables was fully occupied. We could treat six girls at a time. Our annual average fluctuated above the set norm, between 5.2 and 6. We pulled our weight and we chased our sunbeams.

2. Number of Staff

- **Norm**: Number of staff considered essential was set at 10 to 16; less would be rated "best".
- **Comment**: We wanted to test whether fewer personnel would bring staff and juveniles closer to each other and by that open up mutual trust. Institutional staff often called for more consensus building time which meant less time with the juveniles.
- **Result**: "best". In the first year, we averaged 4.5 staff; in the next 2 years, 6.5 staff. This meant that during the 3 year trial period, the average number of staff was 5.8 and during the 10 year period, 7.8. Frossarbo Stables cost much less to run than any other Section 12 home and it indicated that number of staff was not the crucial factor.

3. Attendance

- **Norm**: 10 absconders per year; fewer would be rated "best".
- **Comment**: The juveniles were sentenced to an institution. That they try to run away, particularly at the beginning, is of no surprise. We wanted to assess whether girls who chose to come to Frossarbo Stables would abscond less.
- **Result**: "best". Not once, during the 10 years, did we get to 10 girls absconding in a year. Any absences were in connection with a specific pupil in a specific situation and then, nearly always during the first few weeks. On the fifteenth of December 1987, however, two of the girls walked the six mile long road down to our nearest village and caught a bus from there to the town of Uppsala. This was because one of the girls had a friend who had gone to America and was reported to be living a high life. It was there the girls intended to go. One result of this getaway was a lesson in geography!

4. Failed Leave of Absence

- **Norm**: An average of 10 to 15 leave failures per year was the norm; less than 10 would be rated "best".
- **Comment**: At most Section 12 homes, leave of absence was allowed for 1 to 7 days every second month, depending on how the inmate had responded to the life at the institution. This leave could be seen as a test of how the young person coped. We never spoke of leave of absence at Frossarbo Stables. Every sixth week, each pupil had stable duty. During that time, she had to be at Frossarbo. All other weekends, apart from the first couple of weekends after she arrived, were free. We decided, together with the girl, if she had a good alternative place where she could go, or if she would do better to stay at Frossarbo Stables. Many weekends, the pupils could go home, visit a friend or relative, or go to one of the weekend homes that we sometimes had an opportunity to use.
- **Result**: "best". In spite of having more leave of absence per pupil than was usual in Section 12 homes, we never reached 10 failed leaves in any of the 10 years. At the end of the weekend, the girls always came back, even at the beginning of their stay. As one girl put it: "Of course we had to come back. We were afraid, if we weren't there, one of the other girls would worm her way into our horse's affections."

5. Assistance from a Closed Unit

- **Norm**: Between 4 and 8 requests for assistance from a closed unit per year was the norm; less than 3 per year would be rated "best".
- **Comment**: In certain crisis situations, an open institution may make use of a closed unit to keep the young person in situ for a short time.
- **Result**: "best". During the 10 year period, we used assistance from the closed unit at Hammargården 3 times; twice in connection with the same pupil. Her treatment at Frossarbo Stables had reached a crisis. The girl refused to co-operate and ran away. She was posted as wanted by the police and, when found, she was placed at Hammargården. On these occasions we were very careful to affirm that this had nothing to do with punishment or an attempt to make someone afraid. It was based solely on what treatment was best for her under the circumstances. Was Frossarbo Stables the right place for this girl, or were there, at that stage, better alternatives? The staff at Frossarbo Stables did not make the decision. It was up to the staff at Hammargården to decide, together with the girl. If it was thought that Frossarbo Stables was the best place for the girl, then she would always be welcomed back. It was important that the staff at the closed unit were not cast in the role of "punishing sods", where there would have been no winners, either for the adults or the girls. The girl came back.

6. Period of Treatment (according to choice)

- **Norm**: An average of a one-year treatment period was judged to be evidence that the pupil had co-operated and that the choice of treatment had worked; a choice by the girls to stay more than 1 year would be rated "best"
- **Comment**: At Frossarbo Stables, the youngsters always had the choice of continuing or not.
- **Result**: "best". Seven girls chose to leave. The 40 who remained, did so for periods of treatment at Frossarbo Stables ranging from 18 months to 2 years, a period chosen by each individual herself.

7. Staff Time-out for Re-grouping and Re-starting

- **Norm**: If Frossarbo Stables needed 65 to 100 working days during a 3 year period for re-grouping and re-starting, it would fall within the norm; less than 65 days would be in the range of "best."
- **Comment**: In the Section 12 care system, it often happened that a unit had to re-evaluate, re-organise and re-think. For this, staff time-out was needed.
- **Result**: "best". In the 10 years, Frossarbo staff did not once stop work or demand time to re-group or to re-start.

8. Media Interest

- **Norm**: Up to 2 favourable reports and not more than 4 negative reports during the 3 year period; any additional favourable reports would be in the range of "best".
- **Comment**: Section 12 care continually attracted the attention of the press. During the 10 years, attention was drawn to Frossarbo Stables through articles in the five largest daily newspapers, in weekly magazines and in professional journals covering both psycho-social work and horseracing; in 3 radio and 10 television programmes.
- **Result**: "best". Frossarbo was widely reported in all media and all reports were positive. The articles from the first 3 years were collected in a booklet " The Dawn of a Treatment Home" (Stockholm Läns Landsting 1990).

The 4 FTM Specific Assessment Criteria

1. Education

- **Criteria**: Was passing the first examination (consisting of a written theory test and a practical test) within the first 6 to 8 weeks realistic?
- **Comment**: The written test was concerned with the horse's anatomy. It was made up of 6 different pictures where the girls had to name the various bones and joints. About 50 names had to be remembered. The practical test was to harness the horse to a sulky.
- **Result**: Forty girls succeeded in passing their first examination within the first two months. There were 7 who did not complete the exam. These failures usually occurred during the first weeks and ended in a meeting where, together, the decision to choose Frossarbo Stables was revoked, but where a promise was also given that, should the girl's interest be rekindled at a later date, it would be possible for her to return. The other 40 girls all succeeded in their first examinations, both the theoretical and the practical.

2. Independent Living

- **Criteria**: Would it be realistic to let the girls live by themselves?
- **Comment**: There were 3 dwellings at Frossarbo Stables, 1 very large and 2 smaller ones. The 2 smaller houses were close together. The pupils were to live by themselves in one of the smaller houses and the other was a staff house. It was possible to call out from one house to the other. The choice of the smaller houses for living was to keep things as close to normal and "for real" as possible. Not many live in mansions. So the large building was where we dined together, welcomed and accommodated guests, had conferences, Christmas and summer parties, etc.
- **Results**: During the first 5 years, the girls lived by themselves in their house and did not run away. They got up in the mornings and managed a fairly good level of housecleaning, even if they did not always agree with what was considered an acceptable standard. In spite of their distress diagnosis, they handled the situation well. After 5 years, we decided, together, that an adult should always sleep in the student house, because some pupils were afraid of the dark and there were tendencies to mope, especially when new girls arrived.

3. No Locked Doors

- **Criteria**: Was it practical to allow the girls access to all doors except the office safe where the journals were kept?
- **Comment**: The door to the office safe was the only door that the girls could not unlock. The girls had access to all other doors. In the beginning, girls coming from closed units were both surprised and confused about the unlocked doors, but it soon became natural.
- **Results**: The policy remained in force for 10 years with only one breach. At one stage, food was disappearing from the kitchen. It turned out to be a newcomer who tried to take food with her when visiting home. A reminder ever so important. We like to think that we live in a rich country but we do have poor people and the figure is rising. I have also a sunshine story about no locked doors. For almost 2 years, there was a one hundred kronor note pinned to the notice board in the unlocked office. It was a donation from a guest for hospitality; such donations were placed in a fund for our next trip abroad. However, it was forgotten and just stayed on the notice board. That note was a source of happiness. Whenever I felt low, I just had to look at that 100 kronor note and I was up on my feet again.
-

4. Manage a Working Trip Abroad

- **Criteria**: Could the girls cope with the challenges/temptations of travelling?
- **Comment**: Every year there was an intensive study and racing tour abroad involving staff and all the girls. The travelling included at least 2 trips on ferries where it was possible to obtain cheap alcohol. These trips were to further test the youngsters' strength and the soundness of the Frossarbo Model. There were 10 trips abroad in 10 years. In the first 4 years, we also took horses along for racing. As the trips went further and further afield, we eventually decided to leave the horses at home. Nevertheless, the trips were real horse tours. We travelled in an old rebuild bus, redesigned to sleep in. These were trips with early mornings and late nights. Still, they were trips where everything went well. Everybody joined in the very full programme. Of course, there were tedious times and times when tempers ran short, though nobody ever went over the top.
- **Result**: There was no use of crack or alcohol on the ferries. In fact, on one occasion it was one of the adults who could not cope with the surfeit of alcohol, not the girls. We managed one trip every year, as planned. We visited France and Denmark twice and England, Holland, Germany, Czechoslovakia, Hungary and Italy once each.

Results from Follow up Studies

Two subsequent follow up studies provide further answers.

In 1998-1999 a study of how the girls remembered their horses was undertaken as part of a larger follow study. It found the following results for the 37 girls who had completed 2 examinations or more and stayed an average of 18 months:

1 missing	2,7%
2 relapses	5,4%
3 support programmes (following Frossarbo)	8,1%
31 managing well, having their own homes	83.8%

These are rough figures and they do not specify time. They only say that between 1 and 10 years after leaving Frossarbo there appeared to be substantial therapeutic progress for the girls who had participated for an average of 18 months in FTM (Rapport nr.2/01.SiS; Forsling, 2003:115).

In 2001 Ulrika Bergström reported in her follow up study with Jerzy Sarnecki of girls placed in Section 12 institutions between 1990 and 1995, (SiS project number U-27-3001). Part of the study dealt with the youngsters' experiences during the time they had spent at the institutions. The research focused on, among other things, questions relating to staff, the feeling of safety and certain enforceable measures. In a personal comment about the research, Bergström noted that the girls at Frossarbo Stables differed. During their time at Frossarbo Stables they had experienced total safety and there had been no feelings of compulsion whatsoever.

In summary, the Frossarbo Therapeutic Model was evaluated for a total of 10 years and included a total of 47 girls. Seven girls left but of the 40 who remained, 37 girls completed 2 or more examinations and chose to stay for an average of 18 months. During this time, FTM's performance was assessed as in the "best" category for each of the eight Section 12 standards. The "boundary pushers" (such as the unlocked doors, the many examinations, and the yearly trips abroad), really did push the girls' boundaries, but the positive response and performance by the girls was extra proof of the practicality and beneficial outcomes of the FTM.

Discussion

This 10 year study attempts to show the strength that exists in young people in compulsory care and their determination to live and seize opportunities, in spite of carrying so much that is hard and hurtful.

It is not diagnostic orientated research. I have, however, referred to distress diagnoses (a collective label for all the reports that eventually led to the decision for a compulsory care order, a big and drastic decision) to show how our perceptions of, and relationships with, young people in compulsory care institutions have been fashioned by these reports, including what we think possible and impossible. I have tried to show that some of what we think impossible is, in fact, possible.

Originally the work at Frossarbo Stables was to be evaluated in 1990 and a decision was to be made about its future, whether it would have a place in the Section 12 care system. But by 1990, Frossarbo Stables had already become an integral part of the Section 12 care system because positive media coverage had changed the views of the once negative Conservative and Liberal County Council. We had praise not vilification!

I waited before making my final evaluation because I was spellbound by the relationships between the girls and their horses. I wanted to try to describe the power that seemed to come from the interaction between a horse and a girl. This was the great and important thing that I wanted to explore. It was as if that trustworthy relationship also spilled over to us, the staff. If we pushed too hard, the girl stayed because of her horse and by that gave us a second chance. But there was something more. All my working life had been within the compulsory care system for young people, and now, I had been part of something that worked and gave hope. We had found something that made it possible for us to talk about how clever our young people were and the steps forward they were taking. We no longer needed to be part of the misery that hit the dark headlines: *teenagers getting worse and worse*. Headlines that resulted in more secure units, rather than in more trustworthy methods of treatment.

As Sarnecki (1996) showed, our pupils at Frossarbo Stables were not different from other youngsters in compulsory care. They carried the same heavy distress diagnosis although the specific problems often varied. That is only to be expected. The girls chose to come to Frossarbo Stables, not because of a particular problem, but because they wanted to be with horses. By making it possible for young people to experience meaningful and realistic relationships, they could show a strength that we would never have thought possible. Usual practise is to place youngsters in treatment groups according to their diagnosis. Perhaps we should also consider placing them in meaningful social settings where their own interests play a role and where the adults/therapists can work alongside them on meaningful tasks. Bringing the horse into the arena of compulsory treatment (a fundamental aspect of FTM) seemed to have the power to secure such a quality.

Conclusions

The experiment tried to assess whether it was possible to run an "alternative" institution and still achieve security and therapeutic progress.

The researcher's focus is ideas. The practitioner's focus is bringing ideas to reality. The researcher chooses questions to answer. The practitioner has to face his questions and respond practically, often right away. The researcher needs to convince others. The practitioner needs to convince himself.

A few months after setting up Frossarbo Stables, I thought I had all the answers. I wrote about a stable full of wild roses and thistles with unbending strength, a strength and capacity for life which I saw time and time again during my 10 years at Frossarbo. I saw it grow in a stable with old trotters.

As a researcher, it was important to demonstrate the existence of this strength, a strength so important for the juvenile facing change. As a practitioner, I knew that wild roses and thistles could find soil to grow in a stable and I knew that a hundred kronor note could stay on a notice board for almost two years. Now as a researcher I can support the practitioner. The strength is there. **Bringing the horse into compulsory treatment opens new doors of possibilities and calls for more serious research.**

My work and the evaluation of it convinced me that, too often, we describe youngsters in compulsory care in colours that are too negative and dark. We constantly talk about the need for more secure places. But what is offered at these secure places? Some of the darkness comes from the tint of our own glasses. When youngsters do not want to co-operate, we make the dark darker and the space for light decreases. When a horse does not want to go into a dark horsebox, is there something wrong with the horse? Is it our fault? Or....?

As a researcher and as a practitioner, it is important to show the existence and importance of inner strength, a strength so important for juveniles to feel, if they are to dare to face change.

References

Åkesson, N. (2001): *Socialtjänstenkät*, Ungdom. Allmän SiS-rapport 2001

Andersson, B.(1998). *Ett §12-hem för Flickor*. Omdefinitioner i Ungdomsvården. Rapport nr 2.Statens institutionsstyrelse.

Bergström, U. (2001): Uppföljande Undersökning av Ungdomar som under Åren 1990-1995 Har varit Inskrivna på Särskilda Ungdomshem i Stockholm. SiS projektnummer U-97-3001-1. Personlig kommunikation.

Börjesson, B. (1979): Inre Och Yttre Tvång. Stockholm: Tidens Förlag.Forsling, S. (1974): *Behandlingsundersökningar vid Bärby* 1971-1973. Slutresultat och Sammanfattning. Socialstyrelsen rapport 7.

Forsling, S. (1987): *444 Stockholmsungdomar i kris*. Stockholms läns landsting.

Forsling,S. (1990): *Stall Frossarbo och Pressen*. Ett Behandlingshems Tillblivelse. Stockholms läns landsting.

Forsling, S. (1991): *Blir Värstingen Värre?* Stockholms läns landsting.

Forsling, S. (1992): *Värstingflickor, Finns Dom?* Stockholms läns landsting.

Forsling, S. (2002): *Flickan och Hästen. Bilder från ett hem för Särskild Tillsyn.* Statens Institutionsstyrelse

Forsling, S. (2003): *The Girl and the Horse*. BookArt Productions-Gauteng.

Jones, M. (1968): *Beyond the Therapeutic Community: Social Learning and Social Psychiatry*. Yale University Press.

Miller, A. (1998): *Wege des Lebens. Sieben Geschichten*. Frankfurt am Main: Suhrkamp Verlag. Svensk översättning: *Vägar i Livet: Sju Berättelser*. Wahlström & Widstrand, 1999.

Miller, K. & Harrison, J. In conjunction with Baldwin, R; Dancer, S; Miller, D; O´Brien, J; Pownall, H. & Smart, T.W. (1968): *Care and Training of the Trotter and Pacer*. The United States Trotting Association. Harrisburg: Telegraph Press. Nordin, S. (1959): Spänn banden. Stockholm: Bonniers.

Nordin, S. (1982): *Mitt Liv i Sulkyn*. Bilaga. Travskolan, Stockholm: Guidentryck.

Redl, F. (1965): *Controls from Within*. Free Press.

Runquist, W.(1999): *Brättegården 60 år.* Skol och Behandlingshem. SiS Vänersborg.

Sarnecki, J. (1993): *Uppföljning av § 12-vården i Stockholms län*. Delrapport 5. Stockholms läns landsting.

Sarnecki, J. (1996): *Problemprofiler hos Ungdomar Inskrivna på Särskilda Ungdomshem i Stockholms Län Åren 1990-1994*. Vård av Ungdomar Med Sociala Problem. Red. Armelius B-Å. Et al.. SiS.

Winnicott, D. W. (1971): *Playing and Reality*. London: Tavistock. Svensk översättning: *Lek och Verklighet* (1981). Stockholm: Natur och Kultur.

Article Four

The HEAL Model of Equine-Facilitated Psychotherapy and Learning

Leigh Shambo, MSW, LMHC.
Correspondence to:
Leigh Shambo, MSW, LMHC, Human Equine Alliances for Learning (HEAL)
PO Box 84, Chehalis, WA 98532 USA
Phone: +1 360 266-0778
Email: leigh@humanequinealliance.org

Abstract

The HEAL Model of Equine-Facilitated Psychotherapy and Learning (EFP/L) focuses on bonding with the horse as treatment for Post-Traumatic Stress Disorder (PTSD) and other clinical disorders. The author draws on literature from the fields of neuroscience, attachment and trauma psychology, and animal science to explain the human-horse bond, and how the bond itself helps clients resolve troubled emotional history. Methods are explained including the clinical framework of treatment planning, the facilitator's focus and responsibilities, and the typical progression of clients through beginning, middle and final stages of therapy. A case vignette and excerpts from qualitative data gathered from clients treated with this model are presented. These support the author's hypothesis that the bond with the horse enables the client to form new neural pathways in the limbic (i.e. emotional-social) part of the brain.

Keywords

Equine Facilitated Psychotherapy and Learning – EFP/L; Affective Neuroscience; Emotional Regulation; Limbic System; Post Traumatic Stress Disorder; PTSD.

Introduction

I am Leigh Shambo, a licensed clinical psychotherapist who has used Equine-Facilitated Psychotherapy and Learning (EFP/L) in my private practice for over a decade. A lifelong horsewoman, for many years a professional instructor and trainer, I founded Human Equine Alliances for Learning – HEAL— an organization supporting EFP/L services and research, in 2000. I am the lead author of a HEAL-sponsored study, "Equine-Facilitated Psychotherapy for Trauma-Related Disorders", (Shambo et al., 2010). HEAL also offers professional training for facilitators in EFP/L.

My purpose is to describe the HEAL Model of EFP/L, which we use to address complex Post-Traumatic Stress Disorder (PTSD) and other clinical disorders. Many of our clients have a history of relational breakdown and dysfunction, leaving deficits in emotional regulation and capacity for healthy attachment bonds. In order to address both of these issues, the HEAL Model focuses on the client's connected bond with the horse, as they build a mutually beneficial working relationship. The goal is to build client competencies in key affective (having to do with emotion or emotional expression) domains, and help them translate skills to human environments.

I will introduce this approach with supporting Literature from current scientific fields. The Method section will describe typical HEAL methods as the client progresses from the beginning to the final stages of therapy. The Results section will summarize qualitative findings from our PTSD groups in 2008-2011. The Discussion section explores challenges for the therapist/therapy team, and contrasts the HEAL Model with other models. I will conclude with what I believe to be the important principles which HEAL contributes to the field of EFP/L.

Background

Supporting Literature

How does a connected relationship with the horse help the client heal? The answer lies deep within the mammalian brain, in the brain's limbic system and its body-based partner, the autonomic nervous system (ANS). The limbic system is a set of related structures found in the mid-brain of humans and other mammals. It is located between the primitive, reflex-driven brainstem, and the topmost layer of the brain, the cortex. This "triune brain" is a commonly accepted schematic in neuroscience today (Panksepp, 1998).

Though emotion involves all parts of the triune brain, the limbic system is recognized as the brain's emotional core (Lewis et al., 2000). Limbically-generated affective states are felt throughout the body via the Autonomic Nervous System (ANS). The emotional-social reactivity of the ANS includes near-instantaneous reactions (a pounding heart), and hormonally-based enduring "mood states" (feeling blue). ANS functions include heart rate and respiration, blood pressure and distribution, and muscle tension including facial expression.

Limbic neural pathways develop early, before verbal and logical capacities; they form templates for attachment and belonging throughout life. Trauma, family dysfunction and violence can set the limbic brain and ANS on a permanent "false alarm"; imprinting the mind-body system with hyper- or hypo-arousal, impairing the development and functioning of the reasoning cortex (Rothschild, 2000; Szalavitz et al., 2010).

Moving a person from trauma to healing requires restructuring of emotional response. This must come from experiences that engage and soothe, or "regulate" this sensitive limbic region of the brain. Lewis, et al.'s book "A General Theory of Love" explains the neurological re-wiring that can happen within a bonded relationship. Limbic neuroplasticity – the remodeling of affective neural pathways and responses— requires three stages:

- limbic resonance (a shared empathy that enables depth of understanding);
- limbic regulation (feeling accepted/understood/respected in that relationship), and
- limbic revision (healthier template for future relationships)

(Lewis et al. 2000; Siegel, 2010; Doidge, 2007).

Can the working bond with the horse create such emotionally corrective experiences?

Research by Panksepp (1998) has proven that the limbic system is remarkably similar in humans and other mammals, enabling pair bonding, parenting behaviors, imprinting and enculturation of young, and everyday social community bonds. Horses keenly sense limbic activity and ANS arousal in others. Horses and humans are "highly social" by nature, a cornerstone of our long alliance (Grandin et al., 2009).

Highly social mammals need contact with others, for their own neural regulation. In extreme circumstances the ANS signals fight-or-flight; during times of safety it can 'rest and digest'. For these extremes and all the states in between, social mammals attune to each other— for validation of concerns, reassurance and comfort. Belonging with, and counting on others, is essential for highly social mammals, whose brains rely on the regulating power of relationship (Lewis et al., 2000; Siegel, 2010). Without social connection the brain suffers.

Horses and humans share similarities in social structure as well. Wild horses live in small, family-based bands, sharing a wider range with hundreds of other bands that make up a herd (Grandin, 2009). This helps to explain the vast relational memory that horses have. They readily recognize dozens of clients, sometimes after only one distinct encounter.

Horses are prey animals, reading humans limbically at [physically] wider angles of view, better than humans can read them, with a narrow field of view and an overdeveloped neocortex getting in the way (Grandin, 2005). Key facial expressions are universal to cultures throughout the world. This assures a limbic connection between therapist, client and horse; the client being particularly transparent when limbically connecting with the horse. This seems remarkable, since feelings are often expressed facially in milliseconds and fade almost immediately (Lewis et al., 2000).

Limbic similarity allows a person and a horse to form an evocative and memorable connection. The client experiences emotional regulation within a real-life working bond with the horse. The bond is not sentimental (though it may be full of feeling) nor is it a one-sided fantasy bond. It is a palpable bond of attention and respect which person and horse maintain, and it enables working safely together on challenging, even risky tasks. All tasks with a horse are risky without such a connection!

EFP/L opens the therapist-client relationship in significant ways when the honest and impartial horse becomes a player. Both therapist and client are caught up in the surprising, even mysterious limbic responses from the horses.

The goal of HEAL EFP is restorative limbic exercise which enables the brain, through limbic neural plasticity, to remodel pathways of arousal, emotion and meaning. Novel limbic experiences enable new pathways, enhancing regulation and integration of brainstem, limbic and cortex (Siegel, 2010). Connection regulates our emotions, facilitates meaningful experience, and wires the nervous system. Simply put, "love grows brains" (Perry, 2010: 26). According to therapist Don Lavender (2006: 53), "Horses are magnets for those... who are limbic-connection deprived. The horse's agenda is to join... symbiosis is the order of the day rather than one or the other functioning in isolation." EFP/L "home runs" occur when new behaviors result from this limbic exercise, and are practiced in human environments and relationships.

METHOD

SOUND CLINICAL PRACTICE

The HEAL Model emphasizes the need for sound clinical practice involving:

- **The therapist**: The HEAL Model is designed for the qualified therapist, who may be cross-trained as horse specialist or work closely with a horse specialist in all horse sessions. The therapist develops a treatment plan, guided by knowledge of the client. EFP should make sense as a good pathway toward treatment goals, and may be integrated with other techniques and interventions in the client's plan. The relative proportions of office therapy versus horse therapy time vary with different types of patients as does the relative importance of the horse as an attachment object.
- **The co-workers**: It is especially important with a clinical population to have a good staff:client ratio, and for the staff to have adequate skills with people and with horses. David Young is my horse specialist and co-facilitator. David has an extensive background in horse training, training in various models of EFP, and has an educated interest in human PTSD. Also in attendance at therapy groups, is a case aide from the referring agency who assists the clients with transportation and knows each client personally. Most case aids have an interest in EFP and some experience with horses. The case aide also serves as a liaison between the HEAL team and the clients' primary therapist within the agency. Ricki Martin, the HEAL Operations Manager is also on the premises during group sessions and can assist with a variety of tasks.
- **The herd**: Our herd of horses includes five to seven individuals. They vary in size, temperament, breed and gender. We handle and train our herd to maximize trust and ability to connect with humans by blending three approaches.
    1. **We provide good *limbic care***, which includes good boundaries, fair and strong leadership from the humans, the horses being able to trust in human gentleness and having their needs listened to.
    2. **We use natural horsemanship style approaches** to work within the horse's instincts, overcoming their innate fears of predation and helping them learn to work in partnership with us. This training method is softened by our commitment to hold relationship above performance, and our consistent attention to the handler's – and the horse's – affective state.
    3. **We train each horse on a small repertoire of task-related behaviors** (such as tricks, or number discrimination), using reward-based clicker training. This allows clients to see the horse engage in thinking process.

In the EFP/L sessions the client is introduced to the horses and supported in building a working relationship with one or more horses. This typically continues over a number of sessions – the actual number of sessions depends on the client's needs, goals and progress. EFP sessions can contain a wide range of specific activities: simple observation, quiet time with horses, robust active play in a round pen, riding activities. Clients benefit from grooming, leading the horse, setting simple boundaries and small challenges such as leading the horse past a scary obstacle. Typically we create the emotional building blocks for limbic connection in the early sessions of therapy and more complex activities come later.

Debriefing happens during and after each EFP/L session; for cognitively capable clients additional office sessions can build on the learning from EFP/L. Debriefing helps the client make meaning of the EFP and apply it in human environments. At HEAL we call it a "home run" when we see healthy new behaviors generalize to our clients' human environments. The therapist may use homework, journaling or art, and other interventions to help anchor the learning. The therapist documents sessions and progress and plans for termination when the client is ready.

HEAL's "Tools for the Journey" - Six Keys

The "Six Keys to Relationship" describe six domains of affective and interpersonal functioning. The Keys help the practitioner keep the focus on the limbic interactions of client and horse. Horses' emotional lives are similar to our own. They have six affective domains which clients can observe. These domains are listed and described in Table 1.

Table 1. HEAL's Six Keys to Relationship

Affective Domain (KEY)	Blue Ribbon Neural Circuit; Associated behaviors	Stage of horse relationship in EFP
Body Awareness "The Master Key"	**FEAR** / sense of safety Primed for escape, freeze. Learning to soothe with breath and anchoring thoughts	**<u>Beginning stage of EFP</u>** Self awareness: baselines, inner and outer; Meeting horses and becoming acquainted; Noticing resonance; mutual choosing.
Boundaries	**RAGE** when trapped; Primed to fight. Establish respect by clear boundary.	Ensuring respect and trust; Boundaries to protection and communication; learning to calm the horse. Comfortable together.
Triggers – Defenses (Divided Self)	**PANIC** = separation anxiety; primed for frantic attempts to fit in; fear of rejection	**<u>Middle stage of EFP:</u>** Figuring out dynamics of the relationship; muddling through friction and discomfort. Noticing and responding to disruptions in connection. Hanging out, "just being".
Initiation – Assertiveness (Yin and Yang)	**PLAY** – rough and tumble, dominance hierarchy – who moves whom? Exercises all emotional circuits, but in friendship and safety	Who is leading? Is the leader kind, fair? Balance of assertiveness and sensitivity; Clear and congruent communication. Leading, longeing, language of assertiveness based on driving cues and inner authority.
Imagination (Deer Paths)	**SEEKING** – high drive state; primed to explore, investigate. New neural pathways.	**<u>Culmination stage of EFP</u>** Performance and play based on practiced ability to read and support each other;
Interpersonal (The Social Brain)	**CARE** Circuits – pair & parent-child bonds, social standing; primed to protect and nurture.	Collaborating and co-creating with trust and enjoyment; feelings of belonging and connection, and accomplishment. Home runs – progress in human domains

The Keys enable an integrated understanding of the identified neural circuits for major affective states – circuits which evoke the same archetypal behaviors (i.e. attempts to escape, distress vocalizations) in all mammals including humans. These primitive emotional circuits are designed to drive key survival behaviors. Panksepp (1998: 51-52) dubbed them "Blue Ribbon Emotions", signified by all capital letters.

An Overview of the HEAL EFP/L Method

The HEAL EFP/L horse activities are built around the natural development of a mutually beneficial, working relationship with the horse. Activities range from observing the horses, to approaching, touching, haltering, grooming; synchronous breathing, learning to soothe a nervous horse; practicing boundaries and yielding through leading and groundwork exercises; active free play in a large round pen, using a join-up model softened with a relationship focus; riding, and interactional games and tricks using reward-based training.

The goal is for person and horse to create a limbic bond: curious about each other (resonance), building mutual understanding through challenges (regulation), demonstrating a smooth partnership and attached relationship (limbic revision, for many of our clients). In many cases there is evidence that the bond is quite mutual. This is not surprising because horses form friendships much the way people do – the invisible chemistry of emotional resonance gives EFP much of its magic.

HEAL's horses are recognized as sentient partners with opinions, choices and a sophisticated limbic point of view. They are given as much liberty as possible while, however, recognizing that horse and client are entitled to appropriate structure and direction for safety. We never condone force, but do recognize that assertiveness and awkwardness are part of the developing relationship.

The therapy overall, and each session, are paced to allow exploration of the affective processes between client and horse. There is a saying in horsemanship: "You go faster by going slow." In a therapeutic situation, a horse's reaction is usually slow and there are times when an observer might think nothing is happening between the client and the horse.... But, if you watch carefully, there is!

Case Vignette

HEAL's 3 Stages of EFP – illustrated using selected sessions from a group format therapy program of 7 three hour sessions

Tina[1], 14 years old, is one of five girls referred by a mental health agency for our adolescent PTSD group. She is waif-like and withdrawn.

Beginning stage of EFP – session 1 & session 2

We educate about safety and Key One, asking the girls to gauge and track their "activation" level as they meet our six horses (separated in individual paddocks). Tina's level of anxious arousal is obviously high; but she will not speak and attempts to engage her make her sullen. When the girls meet the horses across a fence (no touching, yet), it is apparent that Beau is drawn to Tina. Tina takes extra time, eye gazing with Beau (who ignored every girl but Tina). Emotional resonance.

By this process of mutual attraction each girl is paired with a horse.

In the second session the arena is full with five inexperienced (and affectively labile) girls handling horses on lead lines. We focus on Key 2, Boundaries. Co-therapist David stays close to Tina and Beau (who could easily knock Tina with a turn of his head). David shows Tina how to "block" when Beau gives a sideways glance she learns to hold her hand up in a signal, "watch out for me." Tina has to watch Beau, too. Beau soon settles beside her.

Middle Stage of EFP – session 4

The girls handle their horses on lead lines out in a field, each girl challenged to hold her horse's attention on the lush, tempting grass. Tina is struggling, but not with Beau. She's mad at every one wearing a human costume. This work on "Not Grazing" is a prelude to an exercise called "Knocking on the Door". We're exploring assertiveness with Key 4, Yin and Yang, and each girl individually will release her horse into the large outdoor round pen, letting it graze at liberty. Then they approach the horse to request its attention, without touching, in order to put the halter on again. The first four girls are successful in commanding (but not forcing) their horse to return to them. Several times, Tina "knocked" and Beau "answered", but as he turned toward her, Tina was already looking away with a shrug. Beau became confused, less responsive. Time started to drag. I felt myself pushing, unwilling to let her "fail", so I backed off, leaving it up to her. Suddenly I felt Tina get mad. She marched toward Beau. His head snapped up, and Tina softened and stepped backward (the "draw"). Beau came toward her with no more thought for the grass.

Tina did not speak to me as the girls boarded the van back to the agency. [In a later conference with her agency therapist Tina revealed her main reasons for being mad at the world and that she was angry with me for pushing. This session and the resulting conference marked a new resolve for Tina in the EFP.]

Completion stage of EFP – session 6 and session 7 (last session)

In the 6th session, the group took a hike with the horses on lead lines. The excitement was contagious, the horses on high alert with "their girls" out in the wide world. Key 7, New Pathways, feels quite literal here! At the apex of our circuit two deer burst from the bushes and ran right through the line of horses, toward the neighboring woods. The horses STARTLED, but stood fast by their girls, including Beau with Tina.

In the last (7th) group session each girl created a dance to music with her horse at liberty in the round pen. It's not a performance, but an act of spontaneous co-creation with the horse as partner. Beau added a move of his own. Trained to bow on command, each time Tina turned toward Beau, he bowed. Tina did not know he could do this nor had she been presented with a cue. Beau was offering simply to please Tina, who could do nothing but bow back. Her peers could not restrain their delighted admiration. The group had become a significant source of safe support for each girl during the summer break from school.

---

[1] Fictitious name – the client agreed to the use of the material on this basis

RESULTS

A replication of the 2006 pilot study was done in 2008 at HEAL (Shambo, et al. 2010) and a thorough discussion of both the quantitative and qualitative data used in it can be found in Zasloff (2009). In addition, in years 2008 through 2011 we collected qualitative data, using a semi-structured written interview from a total of 23 participants in PTSD groups at HEAL. There were 20 female and 3 male participants.[2]

I sorted responses into three categories which reflect the three stages of HEAL's EFP Model, namely:

- feeling **resonance** with another being,
- working things out including the need to **regulate** each other, and
- feelings of a healthy and dynamic bond that lead to **revision** of the old templates.

One response captures all three themes: *"Gypsy and I worked together almost exclusively the entire program. She showed me how to hold and focus my energy, responded to my learning to set boundaries and by the end of the program I've felt that we are partners in this dance of life. The experience began in joy and has ended in a gentle partnership."*

Following are a selection of quotes illustrating each theme

Resonance: Themes of "feeling seen, understood and valued"

"Feeling special."

"So calm, so relaxing, so in tune."

"Calm, worth, loved."

"They [the horses] helped me in feeling that I had meaning."

"Beau was the horse I worked with and he really showed me that I was a 'worthy' person. Cause I have always been told that I was not worthy and would never amount to nothing."

"We bonded. Magic could read me like a book as well as I could read her."

"Was really a bonding experience and felt really the love and centered. I have made good boundaries and friends with Gem."

Regulation: Themes of "working things out"

"To listen to the way my body is feeling and calming myself down. And about personal space and how important it really is."

"Grounding and balancing myself."

"Getting out of my comfort zone."

"This has given me the courage to stand up for myself and defuse panic attacks, set boundaries and bond – with people too."

---

[2] Informed consent to participate in the study was obtained from all participants (and guardian of youth participants). Study design and implementation approved by Lewis and Clark College Human Subjects Research Committee in 2008; the same design was implemented with oversight from HEAL Board of Directors in 2009, 2010 and 2011.

"It has taught me to set good boundaries, and not be scared."

"Gem was able to reflect to me when my energies weren't congruent. She flowed effortlessly when I was "in tune" with my body. I couldn't have done it without her."

"I learned to calm myself before interacting with others. Make sure people understand my personal space and I respect others' space. I learned to trust others as long as they respect my space."

Revision: Themes of "Emotional competency with the horse; skills taken into life"

"The ability to be assertive with something so much larger than myself, to firmly set boundaries and to feel confident within myself. I am more positive about my future and more decisive in my actions than I was previously."

"The assertiveness skills are something that are huge things for me. Now I'm really able to spread my self-preservation wings."

"I am more aware of being grounded in all aspects of my life often times it's the awareness that I'm too much in my head instead of my body but I hope one day I'll be able to feel comfortable in my body at all times."

"Total impact: I have feelings, a voice, and more tools to use."

"I find myself using all I have learned in very stressful situations, and now understand better how to deal with those moments."

"I am more open and honest, communicate and learn the other persons' boundaries; it's a more open way to relate to that person."

Discussion

The HEAL Model calls for a variety of developed clinical skills in the therapist and other team members. Following limbic processes requires the practitioner to "navigate by feeling" and facilitate the client in doing the same. It is often necessary to follow the horse and trust the horse, whose limbic voice will only be revealed when we are able to wait, allow and watch for it. This may feel very vulnerable to therapists used to more structured interventions.

The HEAL Model requires the practitioner to follow and facilitate the client's subtle signs of arousal or avoidance. In short the therapist should be conversant in both bottom-up (somatic) and top down (cognitive) therapeutic methods. It is also helpful to have some familiarity with experiential process work or Gestalt psychotherapy.

Perhaps the biggest conceptual challenge is the understanding of the limbic world of both client and the horse: recognizing in subtle everyday behaviors the concerns of the limbic brain: Who can I count on? With whom do I belong? What's my relationship to you? Do you notice where you end and I begin? Is this relationship safe? Who do I need to look out for, and why? Resolution of these limbic questions for client and horse will lead to satisfaction and clarity in the human-horse relationship. Limbic understanding of the horse may be the invisible quality referred to in the horse world as "feel".

Since HEAL's inception in 2000 I have trained with other organizations, including EAGALA, Epona (my closest affiliation, where I am an Advanced Approved Instructor) and with NARHA (now Path International). These organizations are listed in the References. There are similarities and differences in these models compared with the HEAL model.

I was trained in the EAGALA method in 2000, attending three Level I workshops by 2002. The EAGALA model has its roots in the treatment of adjudicated youth, using a "challenge model" originating in experiential/outdoor education. It has expanded to successfully include a wide array of applications. Some EAGALA exercises direct an unprepared client to navigate novel situations, often to influence the horse to do something. The risks of this approach are too high for many of my clients, whose stress-response systems may already be overwhelmed or on overdrive. Lacking templates for positive relating, these clients often unwittingly put themselves as risk.

There are clients who benefit from being faced with challenging horse exercises early in the therapy; the HEAL Model encourages the challenge strategy when appropriate for the client and horses employed. At HEAL we emphasize that EFP can be challenging and emotionally and physically safe. All clients benefit from the real challenges inherent in the working bond with the horse in EFP/L.

In 2002 I discovered the work of Linda Kohanov and Kathleen Ingram at Epona Equestrian Services. They have been important mentors. Their work has informed the HEAL Model in substantive ways, though we have evolved to make the HEAL Model more suitable for clinical work.

The HEAL Model takes a more sophisticated and nuanced approach to body awareness, advising caution and clinical sophistication in the teaching of body awareness with clinical clients. The capacity for emotional regulation should be developed in tandem with body awareness. Clients should be prepared by development of emotional regulation skills before doing clinical work on trauma memories and somatic effects of trauma. The emphasis on listening to emotions and getting the messages behind them can be inappropriate for emotionally immature clients at the mercy of unregulated or dysregulated feelings.

There are philosophical differences in the EPONA and HEAL models as well. While clients may have spiritual and transpersonal experiences in therapy, these experiences are not the goal of therapy. The HEAL model emphasizes a working bond with horses. We emphasize the challenges of relationship – also the rewards — and we try to ensure that the bridges are built for the home run.

PATH International (formerly NARHA), has a long history serving the therapeutic horsemanship community by providing standards for therapeutic equestrian instruction and facility safety. PATH does not promote a particular model of EFP/L – deferring to the discretion of the therapist who knows their client – but promotes standards of safety and careful risk assessment. HEAL promotes integration and cross-training for EFP/L practitioners.

Conclusion

HEAL defines a therapeutic model for EFP/L, which gives the therapist a framework that maintains a coherent and implicit structure for limbic and relational healing. The HEAL Model is adaptable enough to tailor for different clients' ages and clinical presentations. For some clients with a trauma history, the bond with the horse can offer a limbic experience that they are simply not prepared to have with a human. "When a limbic connection [or lack of it] has established a neural pattern, it takes limbic connection to revise it" (Lewis et al. 2000: 177).

Because the HEAL Model focuses on a working limbic bond with the horse, safety for the human-horse interactions (adequate but not overprotective) is built in. The model focuses on a bond where horse and human look for, and look out for, each other. This is a practical bond of energy and attention. It allows for subjectivity and intensely personal experiences of meaning while encouraging the reality testing that is a function of healthy ego strength.

The director of the public mental health agency, Cascade, which contracts with HEAL for PTSD groups, has called HEAL EFP "cutting edge technology for mental health." (S. Killilay, personal communication, April

19, 2010). This agency is mandated to adhere to "evidence-based" mental health practices. On the strength of our pilot study, ongoing research and the science behind our model, we have continued to serve this agency for four years. The HEAL Model represents a safe and clinically reliable EFP Model for use with PTSD and other mental health disorders.

Acknowledgements: The author extends thanks to Cascade Mental Health Care, in Chehalis WA, and to Human Equine Alliances for Learning for their support of this ongoing study of EFP.

References

Doidge, Norman. (2007). *The Brain That Changes Itself*. New York: Penguin Books.

EAGALA (Equine-assisted Growth and Learning Association). http://www.eagala.org/

Epona Equestrian Services, http://eponaquest.com/

Lavender, Don. (2006). *Equine-Utilised Psychotherapy*. London: Mrunalini Press Limited.

Grandin, Temple and Johnson, Catherine. (2009). *Animals Make Us Human*. Orlando, FL: Houghton-Mifflin Harcourt.

Grandin, Temple and Johnson, Catherine. (2005). *Animals in Translation*. Orlando, FL: Harcourt Books.

Lewis, Thomas, Amini, Fari and Lannon, Richard. (2000). *A General Theory of Love*. New York: Random House, Inc.

Panksepp, Jaak. (1998). *Affective Neuroscience: The Foundations of Human and Animal Emotions*. New York: Oxford University Press.

PATH International (Professional Association for Therapeutic Horsemanship International) http://www.pathintl.org/resources-education/workshops/equine-specialist

Szalavitz, Maia and Perry, Bruce. (2010). *Born for Love: Why Empathy is Essential – and Endangered*. New York: Harper Collins Publishers.

Rothschild, Babette. (2000). *The Body Remembers: The Psychophysiology of Trauma and Trauma Treatment*. New York: Norton and Company.

Shambo, L., Seely S. K., and Vonderfecht, H. (2010). *A Pilot Study on Equine-Facilitated Psychotherapy for Trauma-Related Disorders*, Scientific and Educational Journal of Therapeutic Riding 2010 ISSN 1174-6238.

Siegel, Daniel. (2010). *Mindsight: The New Science of Personal Transformation*. New York: Random House, Inc.

Zasloff, Randy. (2009). *Equine-Facilitated Psychotherapy: An Outcome Study*. Unpublished Masters' Thesis accepted by Lewis & Clark College School of Education and Counseling, Available from author at http://www.findingflowcounseling.com/.

ARTICLE FIVE

The Role and Method of PATH Intl. in Sound Equine-Facilitated Psychotherapy Practice

M Leif Hallberg, MA, LCPC, EFP/EFL Membership Task Force Chair, Acting Communities Coordinator PATH International, Denver, CO
Correspondence to:
Leif Hallberg, MA. LCPC
3214 SE 26th Ave., Portland, OR 97202, USA
Phone: +1 406-600-2004
Email: lhallberg@pathintl.org

ABSTRACT

The work of a PATH Intl. task force offers a unique view on the practice of equine-facilitated psychotherapy. The context for this orientation lies embedded within the broader and long-standing approach of PATH Intl. to equine-assisted activities and therapies and to the organization's general approach to therapeutic work with horses and other equines. This paper outlines the evolution of PATH Intl.'s approach to equine facilitated psychotherapy and summarizes the key elements and requirements of that approach, namely use of practice guidelines, standards, certification and accreditation.

KEYWORDS

Equine-facilitated Psychotherapy; PATH Intl.; Animal Assisted Activities and Therapies, Standards, Guidelines.

Background

The Professional Association of Therapeutic Horsemanship International (PATH Intl.) was founded in 1969 as the North American Riding for the Handicapped Association (NARHA) to promote safe and effective therapeutic horseback riding throughout the United States and Canada[1]. PATH Intl. is headquartered in Denver, Colorado, but volunteers from all across the world contribute to its leadership. PATH Intl. is a non-profit 501 (c) (3) association. The vision and expertise of its membership fuels the volunteer committee structure that in turn determines the direction of the association. Simultaneously, the PATH Intl. Board of Trustees ensures the fiscal viability of the association and sculpts the vision and expertise of the members into a coherent strategic plan. The staff is responsible for the execution of the strategic plan and thus works between the membership and the Board of Trustees, creating timelines and tasks that are divided between staff and volunteers. PATH Intl. has a full time staff of 17 but relies on approximately 200 active volunteers to ensure that the vision and mission of the members and the association is carried out.

The association provides resources such as certification, site accreditation and standards of practice for professionals who provide equine assisted activities and therapies. To date there are over 800 member centers and more than 7000 individual members, of which approximately 5% are international members.[2]

Though PATH Intl. began with a focus on therapeutic horseback riding, the organization has since expanded to include all methods of equine assisted activities and therapies (EAAT). This includes equine-facilitated psychotherapy (a mental health treatment method); therapeutic carriage driving; interactive vaulting, which is similar to gymnastics on horseback; equine-facilitated learning; ground work and stable management; and PATH Intl. veterans and military services, which use a variety of EAAT disciplines specifically to help war veterans and active military personnel. In addition, volunteer committees are working on identifying and refining even more disciplines and activities that might be put to use in the field of EAAT.

PATH Intl. believes there are many ways to provide EAAT, and endorses no one "right" way other than requiring that its members comply with its field-tested standards. The association does not subscribe to any specific method. Rather its approach is to provide ethical guidelines, standards and certification and accreditation processes that enable EAAT professionals to practice safely regardless of their particular and personalized approaches.

Development of Equine-Facilitated Psychotherapy in PATH Intl.

PATH Intl. is in a time of significant growth and change. Probably one of the most important markers of this change occurred in 2010 when the Equine Facilitated Mental Health Association (EFMHA) integrated with what was then NARHA. EFMHA was founded in 1996 as a sub-group of NARHA. The name of the association (North American Riding for the Handicapped) inherently suggested that mental health services and professionals were not included within the body of the organization. Due to this, NARHA's code of ethics, standards, certification, and site accreditation were not crafted to include mental health services or professionals. Therefore, EFMHA created its own code of ethics and its own specialty standards for equine-facilitated psychotherapy. EFMHA also had its own board of directions and its own system of committees which focused on the mental health and learning applications of EAAT. Two boards, two codes of ethics, two sets of standards and the duplication of over 20 committees were redundant. In 2009 the NARHA Board of Trustees asked the question, why are we separate? The idea that mental health issues are innate to daily life has become mainstream. Many of the stereotypes of the past have faded into memory and the growing body of information about the debilitating effects of mental health issues

[1] Information regarding PATH Intl. can be found at www.pathintl.org
[2] Source: PATH Intl. Director of Membership

and the pervasiveness of mental illness in today's society drew the attention of people everywhere. It became clear that mental health was a natural component of any human-equine interaction. Therapeutic riding instructors knew it, administrators knew it, mental health professionals knew it, and so it was time for the association to open its arms.

In this effort, NARHA's Board of Trustees in conjunction with key EFMHA leaders met to outline an integration plan for both entities. As opposed to being a merger, where NARHA would absorb EFMHA, the entire association would have to change in order to integrate two very different organizations into one cohesive association. The name of the association had coincidentally outgrown its roots so the timing of an association name change was fortunate. No longer did "NARHA" clearly represent the vision and scope of the association. Another key component to the integration was the mingling of the two codes of ethics into one united code of ethics for the entire association. Finally, the duplication of the committees needed to be attended to, which led to the strategic dispersal of EFMHA board members and other mental health and learning professionals across all existing NARHA committees. The vision behind this plan was to infuse the entire association with the wealth of knowledge housed within the mental health community which would be a valuable asset to any committee throughout the association.

RESULTS

In order to ensure that the needs of mental health and learning professionals would be attended to throughout the integration, a task force was created specifically to investigate and increase membership benefits. The task force outlined a plan of action which included revising the existing equine-facilitated psychotherapy definition, creating a set of practice guidelines, increasing continuing education opportunities at PATH Intl. events, creating a "How To" manual for professionals interested in starting an equine-facilitated psychotherapy practice and increasing networking and community building opportunities throughout the association. Once the task force got started, it became clear that much of the work it was doing could positively benefit the entire association, and soon its work grew to include investigating membership benefits for all members as well as crafting comprehensive baseline criteria for a potential update to the PATH Intl. certification process.

THE GUIDELINES

The task force recognized EFMHA had long held the NARHA belief that there was no one "right" way to provide equine-facilitated psychotherapy. But, its members also believed PATH Intl. had something unique to offer the EAAT field. At EFMHA's core was the belief that equines are sentient beings capable of partnering with humans to provide a mental health or learning service. This cornerstone provided the task force with fodder to begin designing a set of practice guidelines for equine-facilitated psychotherapy. The task force believed that in order to truly understand the concept of animal sentience, a professional needed to be well educated and trained in the physiology, psychology, communication and behavior of the animal he or she is attempting to "partner" with. This concept aligned with the beliefs of the entire association. PATH Intl. prides itself in requiring those wishing to partner with horses to be competent in their own horsemanship skills. Those certified by the association work hard to ensure that they pass either a mounted or ground-only skills tests which demonstrates a certain level of competency in their relationship with equines. This is a highly acceptable method for therapeutic riding instructors or other educators providing equine interactions. Interestingly, for mental health professionals one of the most common models in existence is to pair a licensed mental health professional with little if any equine knowledge or experience with an "equine specialist". Commonly, the equine specialist provides the safety support for the equine and the client as well as designs the session activities based upon the information provided by the licensed mental health professional. The equine specialist is required to know, and in the case of PATH Intl., demonstrate their horsemanship abilities, but the mental health professional is not required to do so.

This information spurred the curiosity of the task force. In the larger field of mental health, it is understood by practicing professionals that in order to include any new method or specialty area of practice, the professional must achieve competency within that new method.

The National Association of Social Work states:

a) "Social workers should provide services and represent themselves as competent only within the boundaries of their education, training, license, certification, consultation received, supervised experience, or other relevant professional experience.
b) Social workers should provide services in substantive areas or use intervention techniques or approaches that are new to them only after engaging in appropriate study, training, consultation, and supervision from people who are competent in those interventions or techniques.
c) When generally recognized standards do not exist with respect to an emerging area of practice, social workers should exercise careful judgment and take responsible steps (including appropriate education, research, training, consultation, and supervision) to ensure the competence of their work and to protect clients from harm."
(NASW, Revised 2008, Code of Ethics 1.04)

The American Counseling Association and the American Psychological Association's state:

> "Counselors practice in specialty areas new to them only after appropriate education, training, and supervised experience. While developing skills in new specialty areas, counselors take steps to ensure the competence of their work and to protect others from possible harm" (ACA, 2005, Code of Ethics C.2.b and APA, 2002, Code of Ethics 2.01.c).

A PATH Intl. representative contacted the American Counseling Association, the American Psychological Association, the National Association of Social Work, and the National Board for Certified Counselors in an effort to fully understand the implications of these and other ethics that might be associated with providing equine-facilitated psychotherapy. In this process, the task force hoped to gain information which would lead to the creation of a set of guidelines that would align with the ethical codes governing mental health professionals. Of particular interest to the PATH Intl. representative was the question of competency. Was the licensed mental health professional responsible for being trained, educated, and supervised in equine-facilitated psychotherapy? Was equine-facilitated psychotherapy considered a "specialty area of practice" by these associations? Could a licensed mental health professional abdicate responsibility for the physical safety of a client due to lack of expertise, knowledge, and training? The answers began to shape the new PATH Intl. Equine-Facilitated Psychotherapy Guidelines[3].

According to the various associations, "equine therapy" as they termed it, is considered a specialty area of practice and thus falls under the ethical codes which address such areas of practice. They also indicated that all licensed mental health professionals must adhere to their own codes of ethics related to competency.

The National Board for Certified Counselors states:

> "Certified counselors offer only professional services for which they are trained or have supervised experience. No diagnosis, assessment, or treatment should be performed without

---

[3] These guidelines can be accessed at http://www.pathintl.org/images/pdf/resources/EFP-Definition-and-Guidelines.pdf

> prior training or supervision. Certified counselors are responsible for correcting any misrepresentations of their qualifications by others" (NCBB, 2005, A.6).

The American Psychological Association Ethical Principles of Psychologists and Code of Conduct states:

> "Psychologists provide services, teach, and conduct research with populations and in areas only within the boundaries of their competence, based on their education, training, supervised experience, consultation, study, or professional experience" (APA, 2010, 2.01(a)).

And the American Counseling Association states:

> "Counselors practice only within the boundaries of their competence, based on their education, training, supervised experience, state and national professional credentials, and appropriate professional experience... "(ACA, 2005 C.2.a).

Furthermore, the various associations stated that the licensed mental health professional providing "equine therapy" is solely responsible for the emotional and physical safety, dignity, and well-being of the client during any service advertised and provided by the licensed mental health professional and he or she cannot abdicate responsibility due to a lack of expertise, knowledge, and experience.

Finally, PATH Intl. learned that any licensed mental health professional who is certified by the National Board of Certified Counselors can be subject to review and possibly lose their license to practice if they violate any of these codes.

The National Board for Certified Counselors states:

> "The NBCC Code of Ethics provides a minimal ethical standard for the professional behavior of all NBCC certificants. This Code provides an expectation of and assurance for the ethical practice for all who use the professional services of an NBCC certificant. In addition, it serves the purpose of having an enforceable standard for all NBCC certificants and assures those served of some resource in case of a perceived ethical violation" (NBCC, 2005, Preamble).

After thoroughly reviewing the results of the ethics investigation, the task force determined it would be in the best interest of PATH Intl. to ensure that its members were in alignment with these various ethical codes.

The new PATH Intl. Equine-Facilitated Psychotherapy Guidelines (see Attachment) state that the licensed professional must be trained, educated, experienced, and supervised in this method. The task force recognizes that the next step is to craft a standard that outlines exactly what the competency expectations are for mental health professionals providing equine-facilitated psychotherapy. In this effort the task force will evaluate other specialty areas of practice to gain a deeper understanding of the scope and depth of acceptable competency. Until that task is completed and approved by PATH Intl., it is the recommendation of the task force that all licensed mental health professionals evaluate their own competency standards and consider aligning with the most stringent competency standards available to-date in this field.

Two examples of organizations currently requiring competency evaluations are the Certification Board for Equine Interaction Professionals[4] and the Equine Assisted Growth and Learning Association[5]. The Certification Board for Equine Interaction Professionals has begun to set the bar for licensed mental health professionals who are considering adding equine interactions to their practices. The board requires

---

[4] All information regarding the CBEIP certification comes from http://www.cbeip.com/

[5] All information regarding the EAGALA certification process comes from http://eagala.org/Certification_Program

1000 hours of practical experience providing equine-facilitated psychotherapy but only requires 60 hours of specific education and does not indicate how many hours of supervision a professional needs to have in order to be certified (CBEIP, 2011, Certification, "Eligibility Requirements" 2 and 4). The Equine Assisted Growth and Learning Association requires the equine specialist to have achieved 6000 hours of hands-on work with horses and 100 hours of continuing education, but does not require that the licensed mental health professional have any additional supervision or hours of specialty practice related to equine interactions outside of the 6 days of training provided through the association (EAGALA, 2010, Certification Program, "Mental Health Professional" and "Equine Specialist") These two examples are contrasted by the specialty area of play therapy. The Association for Play Therapy[6] states that to add play to traditional talk therapy includes having an additional 2000 hours of practical experience, 150 hours of education directly related to play therapy, and 500 hours of supervision (APT, 2012, Credentials Overview, link to RPT/S PDF).

The Path Intl. task force working on these issues encourages licensed mental health professionals to join it in this philosophy of increasing competency requirements in an effort to promote the safety, wellbeing and dignity of the participants and the equines, and professionalism of the field.

Along with the points that encourage licensed mental health professionals to obtain additional training, supervision, and experience in both horsemanship and equine-facilitated psychotherapy, the guidelines document provides information about other aspects of equine-facilitated psychotherapy practice.

It has been the role of PATH Intl. to create and maintain standards of practice for the field of equine assisted activities and therapies and so the association embraces the task of developing competency standards for equine-facilitated psychotherapy. The task force looks forward to gaining the feedback, ideas, and concerns of all those providing equine interactions for mental health in the hopes of working towards a universal understanding of competency.

Conclusion

PATH Intl. and the equine-facilitated psychotherapy community aim to assist in the development of competency standards and the unification of consumer expectations for licensed mental health professionals. The hope is that through outreach, dialog, and networking, a universal community of professionals can assist in the creation of such exciting and timely work.

References

The American Counseling Association (2005). Code of Ethics. Alexandria, VA. The American Counseling Association. Retrieved from http://www.counseling.org/ethics/feedback/ACA2005Code.pdf

The American Psychological Association (2010).Ethical Principles of Psychologists and Code of Conduct. Washington, DC. The American Psychological Association. Retrieved from http://www.apa.org/ethics/code/index.aspx

The Association for Play Therapy (2012). Retrieved from http://www.a4pt.org/download.cfm?ID=28254

The Certification Board for Equine Interaction Professionals (2011). Retrieved from http://www.cbeip.com/certification.aspx

[6] All information regarding the APT certification process comes from http://www.a4pt.org/ps.credentials.cfm

The Equine Growth and Learning Association (2010). Retrieved from http://eagala.org/Certification_Program

The National Association of Social Workers (2008). Code of Ethics. Washington, DC. . Retrieved from http://www.naswdc.org/pubs/code/code.asp

The National Board for Certified Counselors (2005). Code of Ethics. Greensboro, NC. Retrieved from http://www.nbcc.org/Assets/Ethics/nbcc-codeofethics.pdf

The Professional Association of Therapeutic Horsemanship International (2012). Retrieved from http://www.pathintl.org/about-path-intl/about-path-intl

The Professional Association of Therapeutic Horsemanship International (2012). Retrieved from http://www.pathintl.org/images/pdf/resources/EFP-Definition-and-Guidelines.pdf

ATTACHMENT

PATH INTERNATIONAL
Professional Association of Therapeutic Horsemanship International

Revised Equine Facilitated Psychotherapy Definition and New EFP Guidelines

EFP Definition:

EFP is defined as an interactive process in which a licensed mental health professional working with or as an appropriately credentialed equine professional, partners with suitable equine(s) to address psychotherapy goals set forth by the mental health professional and the client.

Guidelines for Providing Equine Facilitated Psychotherapy:
PATH International's Unique Approach:

Purpose

The purpose of this document is to inform PATH International members, health care professionals, and education professionals about Equine Facilitated Psychotherapy (EFP).

EFP is a popular and emerging service provided by hundreds of PATH International centers and thousands of professionals in the United States and increasingly in other countries around the world. Evolution and growth has resulted in a wide variety of EFP approaches, methods, and programs.

PATH International's unique approach to EFP provides support for a diverse group of professionals serving an increasing range of clients. Services include clinical interventions conducted by licensed mental health professionals, as well as long-term psychotherapy.

The common denominator in all PATH International EFP services is the interaction with and participation by the equine.

The following information helps professionals interested in EFP interpret some of the key characteristics of the service.

EFP is defined as an interactive process in which a licensed mental health professional working with or as an appropriately credentialed equine professional, partners with suitable equine(s) to address psychotherapy goals set forth by the mental health professional and the client.

I. **About EFP**
   a. EFP is an interactive process in which individuals of all ages participate in clinically-structured equine activities to improve and enhance mental, emotional, and social functioning.
   b. Through relationships with equines, humans may recognize maladaptive behaviors. In order to feel safe throughout the process of healing and growth, it is essential that trust

and respect are developed between the human and the equine. Therefore, teaching clients about equines is beneficial in order to provide a deeper understanding and awareness of the equine partners.

c. EFP professionals must obtain a signed "informed consent" from their client. This document describes the professional's services, goals, credentials and philosophy for including equines in psychotherapeutic treatment.
d. EFP differs from Equine Facilitated Learning (EFL) as it is a form of psychotherapy and thus involves the processing of experiences, feelings, and thoughts while EFL involves discussion and educational content. EFP is client driven and the goals may vary where EFL usually has an end goal in mind at the start of the lesson.
e. The goals for EFP range from making the unconscious conscious to treating maladaptive patterns and behaviors caused by mental illness, addiction, trauma, or abuse.
f. Currently insufficient research evidence exists to substantiate best practices or evidence-based practices in EFP. Nonetheless, EFP professionals strive to implement evidence-based interventions when possible and seek to contribute accurate data to the growing research and literature in the field.

II. The Professional

a. EFP is provided by a mental health professional who is appropriately licensed or credentialed in his/her state of practice, is dually trained (licensed mental health professional and equine specialist/therapeutic riding instructor) or is partnering with an equine specialist or therapeutic riding instructor, and skilled or certified to provide EFP.
b. It is the responsibility of the mental health professional to obtain the necessary training, education, supervision, and experience in any specialty area of practice.
c. EFP is developing rapidly and therefore requires continuing education.
d. EFP professionals may choose to provide EFP individually, or may use a team approach (licensed mental health professional, equine specialist, and/or other support staff). This treatment decision is be made by the professional based upon their comfort with the method and the therapeutic needs of their client(s).
e. EFP professionals are trained to interpret the equine/client interactions and provide feedback regarding the equine's behavior, and in answering questions and/or assisting the client in the process of interpreting what the equine is communicating.
f. If the needs of the client/participant exceed the training/practice of the EFP professional, procedures must be in place to refer the client to additional appropriate services.

III. The Client

a. There should be a written treatment plan in place that supports the use of EFP as a clinically appropriate intervention and includes criteria for discharge.
b. Use of EFP as a clinically appropriate intervention should be adapted to the client's goals, personality, and diagnoses. Clients should be screened for precautions and contraindications (see Standards for Accreditation and Certification Manual)
c. The EFP professional should make informed decisions about when a client is suited and emotionally ready for such an interaction and when it would be contraindicated. Continued re-evaluation of the EFP service is critical to ensure appropriate use of the method.
d. The professional is responsible for maintaining HIPPA compliant records.
e. EFP professionals should be educated, familiar, and experienced in understanding equine behavior and the psychological and physiological factors that influence equines.

IV. The Equine

a. There should be a procedure in practice to select appropriate equines for EFP and to periodically reevaluate each equine for suitability.
b. EFP professionals should be highly attuned to and aware of their equine partners. When the equine is responding to stimuli, these reactions and responses can facilitate a therapeutic or learning experience for the client/participant. In this manner, the EFP professional and the equine work together in partnership to facilitate the experience.
c. The welfare of the equine should be assured at all times and the EFP activity must cease in the event that an equine displays unhealthy stress, fear or discomfort. Continued re-evaluation is critical to ensure that both client and equine needs are met.
d. An EFP session should count towards the work day limit for an equine.

V. The EFP Session

a. EFP can be delivered using a variety of approaches including but not limited to; ground work activities, mounted activities, driving, or vaulting.
b. If the EFP professional uses mounted work, he or she must be certified as a PATH Intl. Therapeutic Riding Instructor.
c. If the EFP professional chooses to use only ground work, he or she can be certified only as a PATH Intl. Equine Specialist.
d. If driving or vaulting are utilized, the EFP professional must be certified in those areas.
e. There must be written documentation available to support b, c, and d.

VI. The Equipment

a. The EFP professional should consider the possible psychological impact of using certain equipment with or around clients. Such equipment could include lunge whips and lunge lines, wands, chains, and activities equipment like blindfolds, masks, etc.
b. The EFP professional needs to consider the appropriateness of the equipment for his/her equine partners.

VII. The Location

a. EFP must be provided in a confidential setting that is both physically and emotionally safe for clients.
b. The EFP professional must ensure that the facility's practices align with safety and health guidelines as outlined by PATH Intl. standards.

VIII. Personnel

a. The use of a support team can be helpful to mental health professionals in some situations. Support staff may be certified therapeutic riding instructors, equine specialists, or may have other training and skills that make them appropriate in supporting an EFP provider. The responsibilities of the support team will vary with the needs of each provider, client, and equine.
b. Paid and unpaid personnel involved with EFP sessions should go through orientation training specific for the EFP service.
    i. The training should specifically address confidentiality and basic counseling skills.
    ii. It should address mental health diagnoses that the personnel might encounter as well as potential goals that might be addressed through the EFP service.
    iii. It should cover roles and responsibilities for all personnel.
    iv. There should be a pre and post session briefing/debriefing to ensure that personnel are supported in their efforts.

ARTICLE SIX

Incidents and Injury within the Hippotherapy Milieu: Four Years of Safety Study Data on Risk, Risk Management, and Occurrences

Rebecca Cook OTR, HPCS
Correspondence to:
Rebecca Cook OTR, HPCS
5600 Katz Road, Grass Lake, MI 49240
Phone +1 517-914-0800
Email: becky@rightsteptherapy.com

ABSTRACT

Introduction: It is becoming increasingly popular for therapists to provide treatment incorporating hippotherapy. However equine activities involve inherent risk. The aim of this study was to discover incident and injury occurrences within the hippotherapy milieu. Method: Annual data on risks and injury was collected from hippotherapy providers in each of the years 2007 to 2010 via an internet survey. Results: A total of 425 responses, representing 143,855 hippotherapy sessions, revealed a rate of one injury for every 14,386 hippotherapy sessions (0.007%) with no injury being permanent or disabling. This compares favorably with injury rates for recreational and sporting activities with horses. Identified risk factors included horse behavior, weather conditions, and absence of assistants. Conclusions: Because there are inherent risks associated with hippotherapy, a sound risk management program and skilled expertise beyond traditional academic courses is essential for a safe treatment environment.

KEYWORDS

Hippotherapy; Safety; Risk Management; Incidents; Injury; Occupational Therapy; Physical Therapy; Speech Therapy; Equine Assisted Therapy.

INTRODUCTION

It is becoming increasingly popular for therapists to provide developmental and rehabilitative therapy utilizing horses. Hippotherapy is a therapeutic treatment strategy employed in the United States by occupational therapists, physical therapists, and speech therapists with a strong background in movement analysis and knowledge of the horse. The horse, an essential part of the hippotherapy team, is selected based on temperament and training but while horses in hippotherapy are well-mannered and trained for the job, they like any horse, can be unpredictable.

The introduction of a research review paper by Silver (2002) claims *"Injuries are an inevitable consequence of horse riding."* It concludes that *"Horse riding is a dangerous sport."* US Equine liability laws outline "inherent risk" with equine activity due to the propensity for unpredictable equine reactions (Fershtman, 1996). Accidents can happen in any therapy milieu. However, physical, occupational or speech therapy utilizing a hippotherapy treatment approach involves inherent added risk due to the presence and close involvement of horses. After more than 20 years of hippotherapy in the USA, it is prudent and imperative to analyze trends of safety in the equine milieu for volunteers, therapists and clients.

Bixby Hammett (2001) from the American Medical Equestrian Association (now known as "EMSA", Equestrian Medical Safety Association) believes records of horse-related human accidents should provide a means to learn about prevention. However, few keep records of injuries, and studies seek different information with no agreement of what should be reported. This makes comparison difficult.

In 2002 Markel (a primary liability insurance provider for the equine assisted activities and therapy industry) was contacted by the author regarding incidents and injury. During the phone conversation, the Markel insurance agent declined disclosure of specific numbers but commented that incidence was low. Some data on equine related injuries is available. Strickland (2000) revealed the connection between increased equine speed and increased injury in a Cambridge University study of 1000 riding accident hospital admissions that showed the relationship of one injury for 100 hours of leisure riding, one injury for 5 hours of amateur racing over jumps, and one injury for 1 hour of cross-country eventing. According to Carrillo, et al. (2007) in the United States each year an estimated 30 million people ride horses, resulting in 50,000 hospital emergency room visits (1 visit per 600 riders per year).

Bixby Hammett (2001) says that the National Electronic Injury Surveillance System (NEISS) provides the only national figures available on hospital emergency horse-related injuries but that medical personnel may not know or be interested in the events involved in the injury. Furthermore, not all horse related injuries result in a hospital visit. A study by Christey, et al. (1994) surveyed members of a list provided by a national mail-order riding equipment catalogue and found an overall injury rate of 0.6 per 1000 riding hours for persons younger than 25 years of age with the most common type of horse related injury being soft tissue injury (sprains, bruises, lacerations), then fractures, followed by concussion.

Injury resulting from treatment utilizing horses is not made public through liability insurance companies and it would be difficult to determine the extent of hippotherapy incidents from the NEISS hospital based data system. Therefore an alternative method was explored in a retrospective hippotherapy safety pilot study n=67 (Cook 2007) which became the catalyst for the current multiyear report. The pilot study found an average of one incident/injury for every 14,351 hippotherapy sessions. The pilot survey sessions ranged from 20-minutes to one hour with both a mode and mean average of a 40-minute session length. This information is useful for comparison with other studies and would factor to an incident/ injury rate for hippotherapy of 0.1 per every 1000 hours.

METHODS

The survey tool was designed by an occupational therapist with a clinical specialty in hippotherapy and over 20 years of experience in the field of Equine Assisted Activities and Therapy. The initial pilot safety research (Cook, 2007) was a retrospective study with 45 questions. With the help of an additional therapist, the design for the current study shortened the survey to 18 questions: subject matter on demographics, training, experience and number of hippotherapy sessions offered per year were unchanged; safety related questions were retained but re-worded to describe data referring only to the current survey year (versus from all previous practice years); and questions related to business aspects (fee structures, revenue, expenses, cancellations, therapy times) were removed.

Annual survey data was collected at the end of 2007, 2008, 2009, and 2010 via an internet based professional survey company (Survey Monkey) of active hippotherapy providers on the treatment risks and injury occurrences in the hippotherapy milieu with participation requests announced through the hippotherapy list serve groups and the American Hippotherapy Association (AHA) e-bulletin. Responses were voluntary and anonymous to encourage participation but were also limited to one per internet address to prevent data duplication. In summary:

- The survey consisted of 18 questions: (2) open-ended + (16) forced response.
- Survey questions collected data on credentials, experience, number of hippotherapy treatments per year, therapist training, risk management policies, emergency procedures, occurrence rate and record of human incident or injury.
- Each year represents four separate surveys with relevant information collected over a 2-month period starting in December for the given year and ending in January.
- Content of the questions remained unchanged for all collection years.
- The survey request for participation was open to all active hippotherapy providers and was not sent to a pre-selected number of participants.
- Survey data collection is planned to remain ongoing.

The majority of survey respondents were members of American Hippotherapy Association, Inc. and comparing AHA membership with survey responses among professions for all survey years demonstrates the consistency of a representative sampling of each discipline that completes hippotherapy treatment (Figure 1).

Figure 1: Representative Sample by Discipline

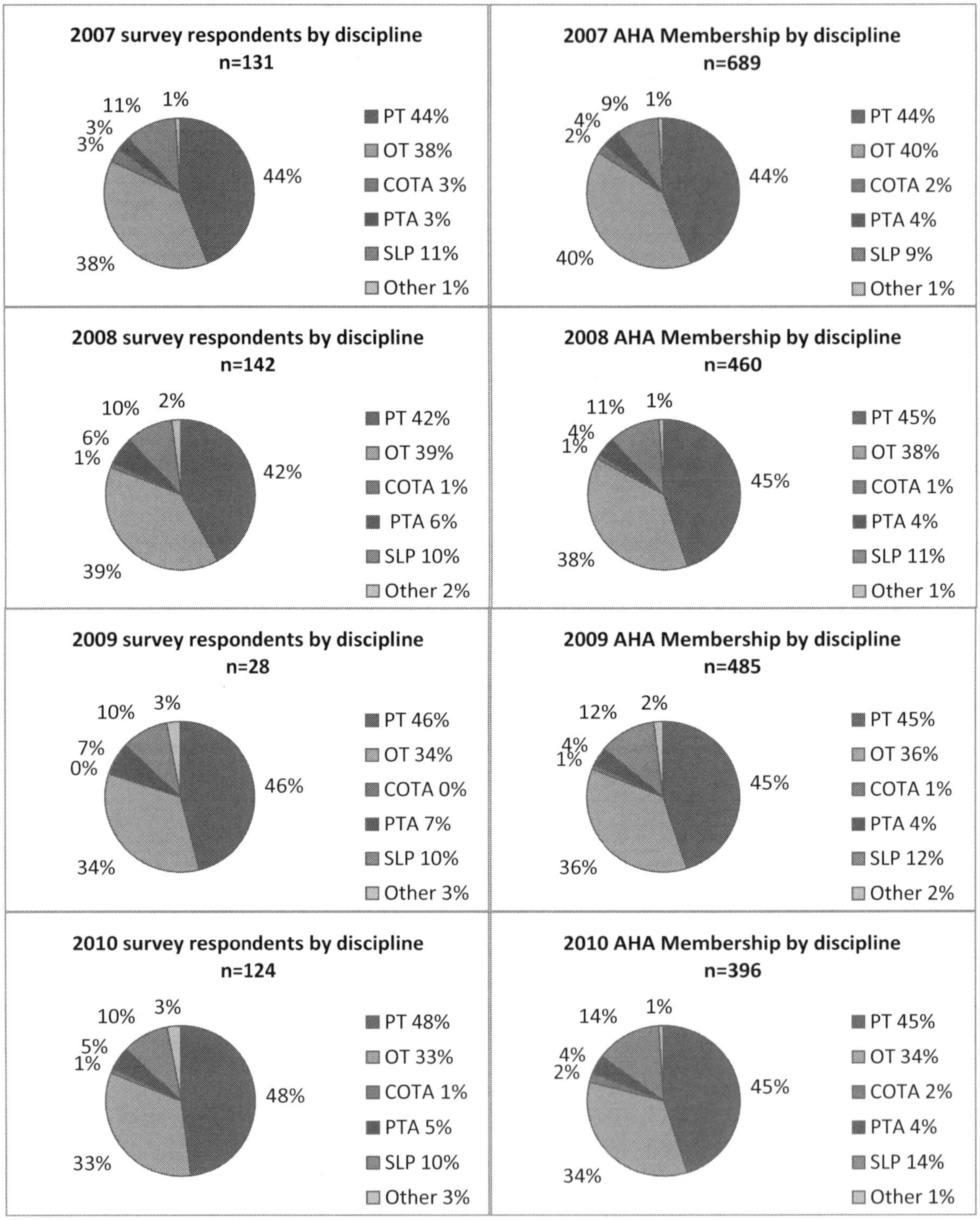

KEY: PT=Physical Therapy, OT=Occupational Therapy, COTA=Certified Occupational Therapy Assistant, PTA=Physical Therapy Assistant, SLP=Speech/Language Pathologist.

It is difficult to come up with an exact survey response rate since a survey invitation was not sent to a specified number of therapists but rather, participation was targeted to hippotherapy providers through list serve groups and the AHA e-bulletin newsletter. However, comparing total annual survey responses to total annual AHA membership numbers can approximate an annual survey participation response rate (Table 1). For many years, a survey's response rate was viewed as an important indicator of survey quality but studies have been conducted that challenge the presumption that a lower response rate means lower survey accuracy. A finding reported by Visser, et al. (1996) showed that surveys with lower response rates (near 20%) yielded more accurate measurements than did surveys with higher response rates (near 60 or 70%). Accordingly, the calculated response rates for this study of 19%, 31%, 6%, and 31% annually (and 21% overall) may yield more accurate measurements than a survey with a higher response rate.

Table 1: Survey Response Rate

	2007 Scores	2008 Scores	2009 Scores	2010 Scores	Comments
AHA annual membership #'s	*689*	*460*	*485*	*396*	*AHA was the primary association to advertise survey participation.*
Total number of survey respondents	131 = 19% response rate as compared with AHA membership #'s	142 = 31% response rate as compared with AHA membership #'s	28 = 6% response rate as compared with AHA membership #'s	124 = 31% response rate as compared with AHA membership #'s	425(survey total) /2030 (AHA total) = 21% calculated as the total overall survey response rate considering all survey years.

Another example from Holbrook et al. (2007) assessed whether lower response rates are associated with less un-weighted demographic representativeness of a sample. By examining the results of 81 national surveys with response rates varying from 5 percent to 54 percent, they found that surveys with much lower response rates were only minimally less accurate. Holbrook's research finding is important for this study since the 2009 response rate of 6 percent was so much lower than the other years. In 2009 exclusively, AHA was running a membership survey simultaneous to the safety survey. The data from 2009 is a small part (n=28) of a larger whole (all survey years combined n=425). There are four years of survey data for evaluation and 2009 conveys information for only one year. However, in spite of low participation, 2009 survey data continues to show a representative sample by profession and survey statistics are comparable to the other survey years, strengthening the validity of the response rate for the entire survey.

Results

Survey demographics including response rate, training, number of hippotherapy sessions; and emergency dismounts, incidents, and injury are presented by survey area in **Table 2** with a comments section to provide clarification and/or analysis of specific data by category.

Table 2: Survey Demographics, Emergency Dismounts, Incidents and Injury

Survey Area	2007 Scores	2008 Scores	2009 Scores	2010 Scores	Comments
Total survey respondents	131	142	28	124	425 total data sets
Training in equine assisted activities/ therapy	98%	99%	100%	98%	Overall results show 99% of survey respondents had additional training and credentialing in the field of equine assisted activities and therapy (hippotherapy).
Number of hippotherapy treatment sessions	**46,656** (mean=356 per therapist)	**48,053** (mean=338 per therapist)	**10,086** (mean=360 per therapist)	**39,060** (mean=315 per therapist)	**143,855** hippotherapy sessions represented in the survey.
Emergency Dismount Rate	**127** (1x per 367 Tx) *49% had **no** Emergency Dismounts*	**148** (1x per 325 Tx) *52% had **no** Emergency Dismounts*	**40** (1x per 252 Tx) *39% had **no** Emergency Dismounts*	**111** (1x per 352 Tx) *56% had **no** Emergency Dismounts*	Emergency dismounts are completed by about ½ of surveyed therapists and the mean average is 1 emergency dismount per every 338 sessions (0.3%).
Incidents - no more than basic First Aid	**17** (1 every 2744 Tx) -8x horse nip, 4x kick, 2x stepped on, 1x bumped, and 2x strain during client assist	**11** (1 every 4368 Tx) -3x horse bite, 4x horse kick, 2x horse bump, 1x stepped on, and 1x strain w/ client assist	**5** (1 every 2017 Tx) -2x horse bite, 1 horse kick, 2x horse spook	**6** (1 every 6510 Tx) -2x horse nip, 2x horse stepped on foot, 2x guided fall w/ no injury	Mean rate of 1 incident per 3,689 sessions (0.03%). Staff and assistants, rather than clients, were most frequently involved with equine incidents.
Injuries – more than basic First Aid	**4** (1 out of 11,664 Tx) -All were session assistants: stepped on x3 and cut x1 and all went to doctor on own	**3** (1 out of 16,018 Tx) -Client fell during mount transfer x2 and client spooked horse x1 w/ 1 hospital stay and all returned to hippotherapy	**1** (1 out of 10,086 Tx) -Therapist sprained Ankle and went to doctor on own	**2** (1 out of 19,530 Tx) -Therapist injured positioning client and one client bruised	Mean of 1 injury per 14,386 hippotherapy sessions (0.007%). In four years, the 10 reported injuries requiring further medical assistance occurred from an equine stepping on a foot, client falls, and therapist body mechanics assisting clients. Six of the ten injury occurrences were with staff. Two client injuries were bruising or occurred from a fall, and two were from mounting.
Person was permanently disabled by an incident or injury?	0	0	0	0	No reported incident or injury was permanent or disabling.
Liability Claims	0	0	0	0	No survey year had any reported liability claims filed.

Common Practice

Workplace details were included in the survey to determine common practice: on average 34% were contract therapists, 33% were employees, and 32% were in private practice, with the remaining 1% volunteering. The majority of therapists (68%) do not own the barn facility or the therapy horses. Common practice from all years of survey respondents showed the top equipment utilized in hippotherapy sessions was the lead rope, halter, and bareback pad as opposed to traditional riding equipment (saddle and bridle).

Risks and Risk Management

When looking at risks associated with hippotherapy, therapists in all survey years identified the top three concerns as:

- horse behavior
- weather conditions, and
- the absence of therapy assistants.

Inherent risk has already been mentioned generally regarding equine reaction but more specifically, weather conditions such as high wind, thunderstorms, heat etc. can heighten an equine fight/flight response and heat, cold, allergy season, barometric pressure changes etc. can also impact client comfort and performance. Hippotherapy is a labor intensive treatment. The hippotherapy team includes the therapist and the horse, the assistance of a horse handler, and depending on the client's needs, an off-side walker may assist with client position or therapy activities. Each team member is essential to the individual client's needs and requires specialized training. A missing team member can impact the risks associated with sessions and the reason why insufficient help was identified as one of the top three concerns.

The most frequent safety strategies and policies enforced during hippotherapy sessions for the survey years were identified by the respondents as:

- staff orientation/training,
- screened and trained therapy horses, and
- cancellation of hippotherapy during extreme weather.

Survey participants were invited in an open-ended comment section to give their opinion on important risk management and safety considerations related to hippotherapy. The themes which emerged from this qualitative data with the greatest frequency of reporting related to the "staff" and "equine" categories, and then to "policies and rules".

Staff: More than 1 out of every 5 comments related to the need for staff to be attentive and aware moment by moment. Next in frequency was having good communication, being able to read the horse, and completing continuing education. Other comments related to hippotherapy staff included: teamwork, orientation to roles, trained, know limitations, adequate number available for sessions, horse knowledgeable, competent, experienced and skilled horse handlers, professional (paid) horse handlers, reliable, consistent, and that the therapist is experienced with horses.

Equine: Comments regarding horses included: screened, well trained for Hippotherapy, complete ongoing conditioning, monitor attitude/demeanor/stress level before and during sessions, easy to read, comfortable work load, safe and reliable, good temperament, various selection to meet client needs, healthy and sound, experienced, receive proper warm-up, tolerant, desensitization training, handled consistently, and trained in the emergency dismount.

Policies and Rules: The third theme occurred in the area of written policies and rules including specifically instructing families and clients on safety rules and risks, checking the facility and environment, practicing safety checks and completing drills including emergency dismounts, utilizing client belts and helmets, and following established equine assisted activity standards.

Emergency Dismounts

The "emergency dismount" is a preventative risk management strategy utilized in hippotherapy to manage client or equine behavior. Staff and horses require training in this procedure and emergency dismounts may occur due to client conduct, seizure activity, horse actions, or weather factors (such as wind gusts or ice sliding off a roof). Emergency dismounts successfully avoid further incident and on average approximately one half of surveyed therapists report completing an emergency dismount at some point during a survey year. The mean rate of emergency dismounts (incident prevention) completed for all survey years was 0.3% or 1 out of every 338 treatment sessions.

Incident and Injury Occurrences

For the purposes of this study, an incident was defined as an occurrence requiring no more than basic first aid, and an injury was defined as requiring treatment beyond basic first aid. The combined average incident rate for all survey years was 0.03% or 1 in every 3,689 treatment sessions; and the combined average rate of injury for all survey years was 0.007% or 1 in every 14,386 treatment sessions. Equine Behavior was identified as the top survey risk and most incidents resulted from horse nips or bites, a horse kick, stepping on a foot, or bumping a person. Staff assisting with the horse on the ground, rather than clients, were most frequently involved with equine incidents and required no more than basic first aid. In four years, the 10 reported injuries requiring medical assistance beyond basic first aid occurred from an equine stepping on a foot, client falls, and therapist body mechanics assisting clients. Six of the ten injury occurrences were with staff. Two client injuries were bruising or occurred from a fall, and two were mounting injuries. In all survey years, no reported injury was permanent or disabling, and none resulted in a liability insurance claim.

Discussion

The top two research questions for any area of practice is whether a particular treatment is safe and if it is effective. The theme of this paper focused on revealing hippotherapy safety data by studying risk management practices and incident occurrences through an ongoing annual survey of hippotherapy practitioners over a four year period. This information is relevant not only to therapists practicing within the field of hippotherapy, but also to clients, parents, farm owners, horse professionals, therapy assistants, volunteers, university fieldwork educators, treatment referral sources, liability insurers and 3rd party payment providers.

In all years the therapists surveyed identified the same principal safety policies for risk management:

1) screened and trained therapy horses;
2) staff and facility orientation/ training; and
3) cancellation of hippotherapy during extreme weather.

From the qualitative aspects of the study, the therapists advise to remain "aware and attentive" every moment during therapy sessions and to "know your horses". Approximately one half of surveyed therapists report completing an emergency dismount (risk management/incident prevention strategy) at some point during a survey year and the mean average rate of emergency dismounts completed for all survey years was 0.3% or 1 out of every 338 treatment sessions. The combined average incident rate (no more than basic first aid) for all survey years was 0.03% or 1 in every 3,689 treatment sessions; and the combined average rate of injury (treatment beyond basic first aid) for all survey years was 0.007% or 1 in every 14,386 treatment sessions, a figure very similar to that reported in the retrospective pilot study (Cook, 2007) of one incident/injury for every 14,351 hippotherapy sessions or 0.1 per 1000 hrs. This rate is significantly lower than the reported hospital injury rate for leisure riders of 1 per 100 hours (Strickland, 2000) or from a surveyed general injury rate of 0.6 per 1000 riding hours for persons younger than 25 years of age. (Christey et al., 1994).

Conclusion

Therapists who choose to practice within a specialty area are ethically bound to obtain additional training and experience in order to demonstrate safety and competence for their treatment of clients. Knowledge and training of therapy equines, being present moment to moment, orientation of therapy staff and clients, cancellation of hippotherapy during extreme weather and other risk management factors are important in maintaining a safe therapy environment. The combination of therapists acquiring additional credentials in hippotherapy and applied risk management strategies in practice may be associated with the low level of incidents and injuries shown in this study. Continuing annual anonymous data collection of risk management and incidents related to hippotherapy treatment will offer integrity and transparency to the community which supports, employs and insures therapy services. These findings may also provide direction for future initiatives in research, education, and practice.

Study Limitations

This study had limitations that are inherent in investigations involving surveys based on willingness to participate and the recall and perspectives of respondents. Invitation for participation was available exclusively online and internet access determined responses. The survey was advertised through associations that are currently supporting training and setting industry standards, and there is an unknown number of physical, occupational, and speech therapists completing hippotherapy who may not be affiliated with these professional organizations; thus it is difficult to determine if therapists who did answer the survey could be identified as providing better risk management and safer hippotherapy sessions than those who are not affiliated with these organizations.

Acknowledgments: A special thank you is extended to Renee Casady PT, MS, HPCS and the American Hippotherapy Association research committee for ongoing support for this safety study. .

References

American Hippotherapy Association (AHA): www.americanhippotherapyassociation.org

Bixby Hammett, D. (2001). National injury data reveal details about accidents. *American Medical Equestrian Association* vol 6, 1.

Carrillo, E.H., Varnagy, D., Bragg, S.M., Levy, J., Riordan, K. (2007). Traumatic injuries associated with horseback riding. *Scand J Surg* 96 (1): 79-82.

Christey, G.L., Nelson, D.E., Rivara, F.P., Smith, S.M., Condie, C. (1994). Horseback riding injuries among children and young adults. *J Fam Pract*. 39 (2):148-52.

Cook, R. (2007). Report on Hippotherapy Survey: Incident/Injury Analysis. *Hippotherapy: The Official Publication of the American Hippotherapy Association* Winter: 10-12.

Fershtman, J. (1996). *Equine Law and Horse Sense*, 2nd Printing. 85-89p. Michigan: Horses & The Law Publishing.

Holbrook, A.L., Krosnick, J.A., and Pfent, A.M. (2007). Response Rates in Surveys by the News Media and Government Contractor Survey Research Firms. Lepkowski, Harris-Kojetin, Lavrakas, Tucker, deLeeuw, Link, Brick, Japec, & Sangster (Eds.), *Advances in Telephone Survey Methodology*. New York: Wiley.

Markel Insurance Company: www.horseinsurance.com

Silver, J.R. (2002). Spinal Injuries resulting from horse riding accidents. *Spinal Cord* 40: 264-271.

Strickland, C. (2000). Equine-Related Human Injuries. *The Horse* Oct 1, 2000, Article 109.

Visser, P.S., Krosnick, J.A., Marquette, J., and Curtin, M. (1996). "Mail Surveys for Election Forecasting? An Evaluation of the Colombia Dispatch Poll". *Public Opinion Quarterly* 60: 181-227.

Article Seven

The Fear Factor: An Important Issue for Safe Equine Assisted Activities & Therapy (EAA/T)

Anita Shkedi MA, Doctoral Candidate (Derby University, UK)
Therapeutic Riding Instructor (Israel National Therapeutic Riding Association –INTRA Israel)
Correspondence to:
Anita Shkedi
Address: 22 Ha Hatzav Tel Mond Israel 40600
Phone: (972) (0) 547205886
Email: anitashkedi@me.com.
Website: www.intra.org.il

Abstract

This is an educational article about the fundamental response to fear, the "flight or fight" mechanism that is inherent in most animal species, and its relevance to therapeutic activities involving horses. By critically examining the hypothesis that the human amygdala's response to fear blocks a person's ability to think, focus, and learn, it is then possible to consider how to prevent it from becoming a powerful detractor in an EAA/T session. Using literature to support this argument, suggestions are made as to how to improve the structure of the EAA/T session so that the challenged rider can feel safe both externally and internally. Challenged riders with various disabilities, including traumatic brain injury (TBI), post traumatic stress disorder (PTSD), and neuro-developmental dysfunctions, phobias and anxiety, are discussed.

Keywords

Fear; Response; Amygdala; Flight or Fight; Amygdala Response.

Introduction

Fear exaggerates and distorts our thinking as we narrow our focus to those things that can harm us. It becomes the lens through which we see the world. Fear is "hard-wired" into our brains and represents a genetic wisdom designed to protect us from bodily harm. It activates a survival system in our bodies that immediately bypasses our rational mind and moves us into attack mode. This state of alertness causes us to perceive almost everything in our world as a possible threat to our survival. We tend to see everyone and everything as a possible enemy (Neimark, 2011).

As such, fear must be considered a serious barrier to a successful EAA/T program as it is a powerful detractor in any challenged rider's equine assisted activities/therapy (EAA/T) session, as well as other environments. Because fear puts us in survival mode, where the rational mind is disengaged, it becomes impossible to cultivate positive attitudes when our consciousness is only focused on fear. In this state of mind it is impossible to make clear choices, as we are overwhelmed with excessive stress, *"our life becomes a series of short-term emergencies. We lose the ability to relax and enjoy the moment".* (Neimark, 2011: 2) We become emotionally insolvent, powerless, unable to cope and problem solve, thus affecting all aspects of life's every day activities, for instance communicating with family friends and colleagues (Shalev, 2002, 2008; Joseph, 2006).

During an EAA/T session, challenged riders are easily frightened when they perceive a threat and feel that they are in a situation in the riding arena that is unsafe or unsure. It is our obligation to try to prevent this from happening by making sure the environment and the equestrian activity is safe and achievable. When a challenged rider perceives a threat during an EAA/T session, his/her brain will immediately activate the fear mechanism and in that moment the brain cautions the challenged rider to be safe ... that they should not ride. This "shut down" action prevents the challenged rider from making any progress, whether it is physical, psychological, educational, social or emotional. They cannot absorb, access or attain and act upon any new information given at that moment in time.

By understanding what triggers fear, therapists can provide a safer environment for riders with disabilities such as Post Traumatic Stress Disorder (PTSD), Traumatic Brain Injury (TBI), neuro-developmental dysfunctions, as well as many children and adults at risk from various physical, psychological social and emotional disabilities.

What Is Fear?

Fear is one of the most human biological and emotional responses. It is programmed into the nervous system and works instinctively. From fetal life, we are equipped with the survival instincts necessary to respond to fear when we sense danger, feel pain or feel unsafe. Fear is felt as a distressing negative sensation and is always induced by a perceived threat (Shalev et al., 1996; Shalev, 2002, 2008; Kidd & Kidd, 1972; Du et al., 2008; Ballenger et al., 2004; Peri et al., 2000).

In extreme cases of fear a person can freeze or adopt the paralyzed position. For example, when working as a health visitor in the UK, I observed a baby, aged ten months, adopt the freeze position with eyes dilated, head held up and trying not to move a muscle when she heard footsteps coming towards her cot. At that young age, she had been conditioned that if she were caught crying, a severe punishment would follow.

It should be remembered that fear is also a protection mechanism, alerting one to danger and preparing one to deal with it. Feeling afraid is very natural and helpful. Fear can be a warning signal that cautions one to be careful.

When danger is perceived by our senses the amygdalae react immediately. The amygdalae are a pair of almond-shaped structures made up of a collection of nuclei, which sit within the medial temporal lobes of the brain. The structures form part of the limbic system and are involved in emotional processing and memory. The amygdalae are implicated in the fear response known as "fight or flight" which is a mechanism common to both human and animals. The amygdala response is supported by the hypothalamus that activates a sequence of nerve cell transmissions that release and then fire chemicals, norepinephrine and cortisol, (Joseph, 2006) into our bloodstream and the brain. Consequently, our respiration rate is increased and blood is directed to the muscles and limbs, which require the extra nourishment for running and fighting. Awareness intensifies, pupils dilate, sight is sharpened, pulses quicken and the perception of pain diminishes. The immune system mobilizes with increased activation. Physical and psychological changes in the environment are scanned and searched. This fearful state of hyper-arousal, like all emotions, can be mild, medium, or intense, depending on the situation and the person. A feeling of fear can be brief or it can last for a long time.

When a riding situation triggers fear, it immediately creates an unsafe, dangerous situation. As fear prevents the rider from having any rational thought, attention or awareness, s/he is unable to cope with the horse and its environment. It is imperative that the EAA/T practitioner considers this fact, even before the rider comes to the session.

What Controls the Fear Response?

The amygdala, as a critical processor for fear and fear learning has the ability to affect an **immediate response** to fear. This immediate response is experiential and based upon what is perceived as dangerous by all five senses (sight, touch, hearing, smell, and taste). The amygdala has the ability to respond in a split-second as it receives information coming from the environment via the shortest possible route to the limbic brain. The amgydala responds and sends immediate information to the brain as a means of ensuring survival. An example of this type of experiential response occurs when someone touches a cup of coffee that is too hot to handle. The skin sensors relay the information through the peripheral nervous system to the central nervous system. The brain reacts by relaying a message to the fingers, "telling" the fingers to set the cup down.

The amygdala is the most important component of the limbic brain and has the greatest influence on emotional responses, consequently influencing all of the brains activities. This means that there is a fear element to almost everything that one does, sees, thinks or feels. It is directly linked with the hippocampus enabling the conversion of incoming information (sensory input) into short-term memory. This short-term memory moves through working memory, where it can either be discarded, or continued to be stored in the long-term memory bank.

Diagram 1. The Brain

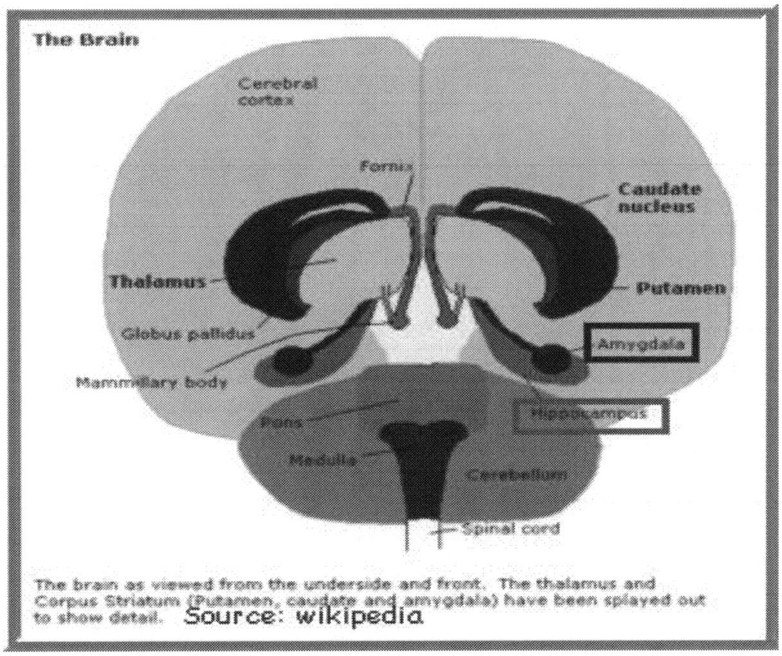

An immediate response to fear begins with arousal. (Shalev, 2002, 2008; Milad et al., 1997; Shalev et al., 1992; Peri et al., 2002)

At a riding centre, fear can be aroused and assuaged by the following situation: The rider climbs up and sits on a horse for the first time (or perhaps after a previous bad equestrian experience). The amygdala is alerted. The horse begins to move quickly, alerting the amygdala even more. Fear takes over – the pupils enlarge, breathing and heartbeat quicken and the feeling of sweating occurs. In many instances an immediate empathetic response from practitioner and horse, for example hearing the reassuring voice of the practitioner and feeling the horse slow down/provide balance can help the rider feel trust, that everything is okay, there is nothing to be afraid of and the fear response regresses. The rider's heartbeat drops to normal levels, as s/he is reassured that there is no danger and the hyper-alertness diminishes in intensity. Although this series of natural occurrences happens, in most cases fear does not disappear so quickly and as easily. It normally takes carefully selected reassurance from both practitioner and the horse to gradually reduce fear. Sometimes it can take days or weeks. The practitioner chooses activities for the rider that emphasize trust, increased understanding and knowledge of the horse and its movement to help overcome the fear factor. If the practitioner allows the rider to choose their own horse and make their own decision as to when s/he mounts the horse again, as well as spending time brushing and grooming and walking beside the horse the negative effects of fear will be greatly reduced.

When a practitioner fails to notice how the rider feels and/or the rider's needs, the following type of scenario can develop: The pressured practitioner says, "Hurry up, let's get moving", and encourages the horse to set off before the rider is ready. The horse starts to feel out of control, the rider feels insecure and unbalanced. The practitioner shouts, "Hold the reins and sit down". Confusion and the inability to understand the command set in. The amygdala is now in full response and is sending information on the fast tract back via the hippocampus (Shalev, 2008; Sanders et al., 2003) to the prefrontal cortex for action, making the rider feel like s/he has reached the point of no return. Nothing can be done now to stop that terrifying feeling and the rider is convinced that a fall and pain will follow.

The situation is finally resolved, but the rider is left with that memory in various separate pictures. The horse, the mounting, the smell of the horse, the reins, the fast movement of the horse and the shouting of the practitioner – in the future every one of these memory pictures can trigger fears and prevent the rider from learning how to ride and enjoy the pleasures of horseback riding. The memory of mounting the horse, hearing the practitioner shouting, or feeling the horse moving fast continues to give the rider a feeling of fear and being out of control.

Kidd & Kidd (1972) confirmed this "Pavlovian" (Sanders et al., 2003) response when they carried out an experiment with rats. After receiving electric shocks in a white box, the rats learned to run to a black box to escape the pain. Following the shock sessions, the rats were placed in the white box that had become harmless. Regardless of this fact they still ran to the black box for safety. However, it can be argued that this experiment shows that one learns by fear, as the rats ran from the white box even after the electricity was turned off. The human amygdala also learns fear conditioning. For example, one would not put their hand in a fire, due to experiential knowledge, as well as biological and genetic experience.

Following the immediate response to fear, the amygdala is in control of a second type of response called the "presumptive response" which interprets the stimuli, and gives meaning to them. For example, a presumptive response might occur if someone was driving a car at 40 mph and a pedestrian stepped off the curb into the road. The driver observes the pedestrian (experiential response) and then, fearful of hitting him/her, slams on the brakes (presumptive response). The presumptive response can only occur after the brain has fully processed the sensory input, and the perceived threat.

Many riders with traumatic brain injury (TBI) are unable to make the presumptive response. As a result of their injuries they have damaged that fast-track pathway to the limbic brain and the amygdala, and therefore are unable to experience such a response. A TBI sufferer's brain often cannot react because the damaged area of the brain disrupts the process. For example, a rider with TBI may be riding in the arena on a windy day when suddenly a plastic bag is blown into the horse's path. The practitioner asks the rider to halt, but although the rider also sees the bag blowing through the air (experiential response), s/he is unable to continue the whole brain process, to perceive the blowing bag as a threat. Consequently, lacking the ability to make the connection, s/he does not come to a halt or make a presumptive response (Shkedi, 2011, in press).

How does Fear Interfere with Cognition?

Fear is present in every aspect of thinking and it is incorrect to dismiss any part of the process, even for hard-core cognitive learning like mathematics. Students studying mathematics are also influenced by the amygdala response. Some argue that emotional responses, like fear perceived from external circumstances, overshadow all cognitive performances. Schloglmann suggests that when a student of average ability sees a teacher writing an equation on the blackboard, which s/he doesn't understand, it appears as a *"set of signs with no structure – immediately the student's fear intensifies, preventing a successful learning process"*. According to Schloglmann, *"Sensory information characterising the situation for that student is influenced by the dynamic representations that comprise the starting point of the observation of the situation"*. He continues to suggest that the *"activation process that arises because of the sensory information constituting the new dynamic representations with respect to the situation is mostly not controlled by the conscious self"* (Schloglmann, 2006:14). Because the set of signs on the black board are perceived as a threat, immediately the amygdala starts to respond to the frightening situation, preventing the student from learning.

Phelps (2006) holds the belief that emotion and cognition are intertwined and are linked always to the amygdala response. She argues that humans, possessing acquired language, can listen to someone passing on information that would prevent a fearful situation. For instance, a report of a dangerous horse in a field, can produce an amygdala response, a warning not go into the field to prevent an attack by that horse.

How Can We Leverage The Horse To Prevent Or Extinguish Fear Over Time?

The brain can inhibit fear, but does not extinguish fear or frightening experiences. By understanding the mechanism of fear, the horse can be leveraged as a technique to actively inhibit fear and frightening experiences. According to Shalev (2002, 2008), a sense of fear is usually re-triggered by a memory (smell, picture, situation) linked to some frightening experience in the past. Consequently, the patient, after responding to a conditioned stimulus (CS), spontaneously recovers with the passage of time. This indicates that extinction does not erase the conditioned memory, but is a form of (active) inhibition. The brain (prefrontal cortex/hippocampus) working closely with the amygdala, learns how to cope with the conditioned memory and suppress the emotional fear reflex.

Diagram 2. The Limbic System

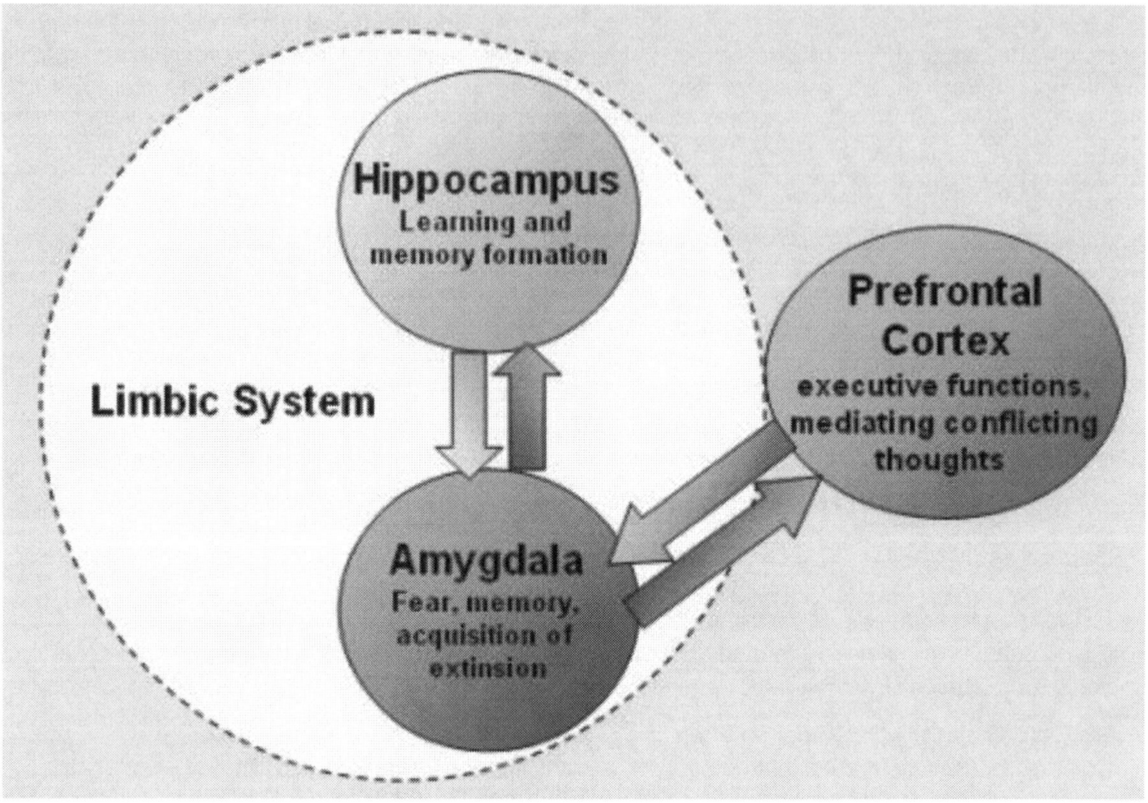

Arieh Shalev, (2008) taken from notes of a reported lecture

IL (Pseudonym) Case Notes, (used with parental permission) Shkedi (2011) INTRA

IL is a small child who was born with a severe form of genetic ataxia and came to the riding centre for therapy. He suffered from a lack of balance, co-ordination and motor planning skill. These disabilities caused him to often fall and hurt himself. Consequently, he had developed a fear of falling and avoided most situations where he might fall. For example, he would only stand alone if he were on a soft surface like rubber matting or carpet, but never on a hard surface like concrete. On the horse, he was alerted to the fear of falling-danger when he attempted to trot, because at that pace he lost his balance and control of his body, head and legs. Fear was immediately aroused and an overwhelming sense of panic, because he thought he was going to fall. Each slight bounce on the pony, after one or two steps, produced a new sensation of danger, and a feeling of being out of control. This episodic fear of falling triggered the amygdala and the fear response. To avoid this situation IL made it known that he did not want to trot. He preferred avoidance rather than confrontation. The result was that instead of overcoming fear, he felt even worse about trotting. To cope with this situation, the technique of CS, a few steps of trot, with long walk was introduced. Not using the trot in every session meant that IL would not be anticipating fearful moments. I have been working on the premise that during the coming year, after CS at very short intervals, IL's prefrontal cortex will start to calm the situation down by telling his amygdala mechanism that "it's Okay".

How Do People Overcome Fear Through EAA/T?

Fear can be healthy if it cautions a person to stay safe around something that could be dangerous. Spending time in EAA/T to try to prevent and/or extinguish fear is worth it, as fear can be changed into empowerment. If a rider with a severe impairment has the opportunity to ride routinely in a safe environment and the horseback-riding session is repetitious, the challenged rider will gain self-confidence, self-esteem and a sense of purpose. If practitioners achieve the best environment for a fearful person, the rider will be able to improve cognitively, emotionally and socially by – as Kidd & Kidd found – removal or prevention of a conflict induced stimulus through reinforced learning. (Kidd & Kidd, 1972)

Fear protects one by being an emotion that is pre-programmed as an instinctual response to potential danger, and survival. It is very important that our bodies react before our conscious mind has even registered the danger. According to Ballenger et al., a neuro-endocrine change involving the hypothalamic-pituitary-adrenal axis, releases enormous amounts of norepinephrine, serotonin, glutamate, y-aminobutyric acid and endogenous opioids into the blood stream, reinforcing the amygdala's response to activate the body for flight or fight (Ballenger et al., 2004).

Fear is not only an instant reaction, as it does relate to future events, such as worsening of a situation or continuation of a situation that is unacceptable. Fear provides the brain with an amplified emotional response that over time is registered by the amygdala, which builds a salient landscape, a map detailing the emotional significance of everything, in order to respond more appropriately in the future.

Fear, whatever its source, can become a controlling factor in a person's life as described in the following example of a person with PTSD:

A mother who lost her daughter after complications of a simple (and unnecessary) intervention, continued to live in the past, persistently re-experiencing the traumatic event. Her daughter died of sepsis, and from that moment in time her mother remembers the death as the most frightful moment: "*I entered the door, my hand still holding the knob. There she lay staring with pupils so dilated that her irises were no longer visible. Death was inevitably approaching. I wanted to scream for help, but there were no doctors present and the nurses were all running around. I could do nothing about it*" (Notes taken from a lecture given by Shalev, 2008).

Shalev suggested that this mother felt desperate and helpless while facing the inevitable and that in PTSD patients the normal extinction mechanisms do not work, as they remain in a state of arousal. He also argues that the mere re-iteration of the traumatic event does not help the patient. If the patient is in fear, it does not help to bring her to the event again and then to leave it as soon as the patient gets frightened. This only reinforces fear. What he says is that the patient should be taught to associate the event with other positive events through psychotherapy. However, he adds that trust, empathy, friendship can help (Shalev, 2008).

During the past year at INTRA, veterans with PTSD have received EAA/T. In an atmosphere of security, friendship and empathy many of the veterans have found a way to cope with their fears. They have found that the horse environment changes their reality. The environment is less toxic, noisy or hostile than the outside world. The centre's atmosphere is emotionally safe as they mix with horses, friends and people who genuinely care for their well-being. Building a human – horse relationship for many of these riders gives them a sense of purpose, of being more worthwhile, and more spiritual.

An on-going study at INTRA examining EAA/T as an effective therapy for veterans with PTSD, recorded changes in the riders' perceptions of reality, mental perspectives, attitudes, beliefs and emotional reactions to the fearful events that happened in the past. Leveraging the properties of the horse, the practitioner can set up a dialogue between horse, practitioner and rider, rebuilding trust, communication skills, reducing emotional numbness and avoidance, lack of sleep and more. When EAA/T gives a voice to the rider who had lost any feelings of self worth, it does not change the physical reality or historical experience of the rider with PTSD, but it can alter the rider's self perception and recent interpretation of feelings and responses. It can be argued, that during the EAA/T session even the hyper-arousal state is lowered, as has been reported by many veterans who began to feel safe in the atmosphere of the centre and the security of the horse. EAA/T alters the challenged rider's perception of reality – viewing the difficulties of life as events that made him stronger and more loving. In the Buddhist tradition, this is referred to as developing a supple mind (Neimark, 2011: 4).

EAA/T as a Physical Exercise

"Perhaps the simplest, best way to turn down the activity of our fight or flight response is by physical exercise. Remember that the natural conclusion of fight or flight is vigorous physical activity. When we exercise, we metabolize excessive stress hormones – restoring our body and mind to a calmer, more relaxed state" (Neimark, 2011: 4).

Horseback riding reduces stress and the flight/fight response by effectively metabolizing and preventing excessive buildup of stress hormones. Physical abilities in particular, can be improved through the horse's rhythmic three-dimensional movement that simulates the pattern of the human walk. When sitting astride a walking horse, its rhythmic movement improves posture, balance, equilibrium, body and spatial awareness, and co-ordination. At the pace of a walk the challenged rider's pulse raises to 120 beats per minute, as it is an aerobic experience (Riedi, 1988).

How Fear Differs From Anxiety?

Fear can be distinguished from the related emotional state of anxiety, which typically occurs without any certain, or immediate external threat. Anxiety is the result of threats which are perceived to be uncontrollable or unavoidable. Many challenged riders with social and emotional problems suffer from anxiety. Many at-risk children from neglected backgrounds suffer from anxieties which interfere with their ability to learn and engage in normal activities. EAA/T practitioners have the potential to mitigate fear and model mastery and reflective responses when they use the unique ability of the horse to mirror fear and anxiety, as a corrective response similar to soothing, reassuring and structuring the response to alleviate apprehension and nervousness which are lasting effects of neglect.

Phobias

Some challenged riders suffer from both anxiety and phobias. Du et al. suggest that a phobia is a fear that is out of proportion to the potential danger. Direct conditioning, vicarious learning, and negative information provision and biological preparedness can acquire a phobia (Du et al., 2008). In the case of horseback riding the phobia or the danger feels real, because the fear is so very strong. When people have phobias, they worry, dread, feel upset by, and avoid the things or situations they fear, because the physical sensations can be so intense. Therefore, a phobia can interfere with normal activities.

Sometimes, challenged riders are extremely afraid of a horse, worry or dread making contact because of what they have been told about the horse from cultural beliefs and stories or simply, because the horse is large, strange and seems uncontrollable. They have been conditioned by environmental, genetic and/or biological pathways to feel afraid of horses (Duet et al., 2008). In extreme phobic cases the person will not come to the center, because they imagine that the horse will do something to them.

What Causes Phobias?

Sometimes phobias develop after a frightening experience with a particular thing or situation. The amygdala keeps track of all experiences that trigger strong emotions and fear. Once something triggers a strong fear reaction, the amygdala warns the person by triggering a fear reaction every time s/he encounters or even thinks about it. Having a phobia is not a sign of weakness or immaturity. It is a response the brain has learned in an attempt to protect the person. It is as if the brain's alert system triggers a false alarm, generating intense fear that is out of proportion to the situation. Because the fear signal is so intense, the person is convinced the danger is greater than it actually is.

Overcoming Phobias

People can learn to overcome phobias by gradually facing their fears. This is not an easy task, especially around horses. EAA/T practitioners have to be aware that it takes willingness and bravery and a lot of support to overcome phobias. The practitioner has to guide the person through the process. Over a period of time the practitioner shows a rider how to approach and come close to a horse, brush it, and walk beside it, before ever thinking of mounting it. The practitioner must try to reduce the fear by talking about the horse and its needs to the rider. When riders find that what they fear does not actually turn out to be true, it can be a great relief and they begin to enjoy being with horses.

During a riding session the practitioner can teach relaxation exercises, specific ways of breathing, muscle relaxation training, or soothing self-talk. These strategies can help the rider feel comfortable and bold enough to face the fears on their list. When s/he finds that they can move in time with the three dimensional movement of the walking horse, they often find that their fear diminishes and is transformed into empowerment.

Discussion

Fear is an inherited instinctive response that has been with us since the beginning of time (Shkedi, 2012 in press). Fear is a type of behaviour that can be engaged with or without the aid of previous experience. The activation of fear in humans is as constant as it is with animals (Phelps, 2006 cites La Bar et al., 1998), and therefore it is beneficial to remember that horses also have an immediate flight/fight response to fear – somewhat like our own. EAA/T practitioners should always remind themselves of this fact, along with the truth that fear can be a group reaction and a social response that is also common to both humans and horses. An unsafe situation can arise in a group session in the arena, where a rider's fear may become contagious, causing the other riders and horses to react fearfully as they capture his or her fear or uncertainty.

The mechanism of fear is multifaceted and the biological response to extreme events is complex. It involves mechanisms related to aspects of survival, learning, memory formation, loss, and socially modulated re-adjustment and adaptation (Shalev, 2008). Fear is an acute response to a perceived threat or to an anxiety disorder like PTSD, or it can be closely associated with disorders like phobias.

EAA/T practitioners must be aware that emotions and cognition are always intrinsically linked and therefore it cannot be said that we can learn without emotional attachment.

EAA/T practitioners must also be aware that fear can interfere with learning as it has the ability to bypass attention and awareness. Consequently, it is incorrect to think that capturing a rider's attention is enough to prevent a traumatic event/accident. Prevention has to start before the challenged rider reaches the horse environment. The EAA/T practitioner has to consider not only the needs of the challenged riders, their emotions and cognition, but also the responsibility for the horse and for making the experience safe and nurturing.

It has been said that it is possible to learn from fear. For instance, the rats that ran from the white box to the black box learned to run to the black box for safety even after the electricity was turned off and they received no shocks in the white box (Kidd & Kidd, 1972; Phelps, 2006). This is not the way for EAA/T practitioners. Riders overcome horseback riding fear by being given the chance to learn about horseback riding gradually. EAA/T practitioners must make it an important part of their therapy session to reassure and relax the rider so that they can have an enjoyable ride. The practitioner must make sure that the rider can understand what is expected of them while they are in the company of a horse. This means that the challenged rider needs to be taught how to recognize and cope with unsafe situations as part of learning equestrian skills. If the practitioner gradually and safely supports the rider by showing how to cope with the unfamiliar experience of being with horses and horseback riding, the rider will feel less vulnerable and unsure. In doing so, the practitioner sets the scene for further learning to take place.

Conclusion

To be able to create the most effective, safe and enjoyable EAA/T program for riders, practitioners must have a good understanding of what fear is, an appreciation of what it means in the lives of riders and how best to handle it in a safe and positive way.

Acknowledgement: Sources of Support from Professor Arieh Shalev, Professor of Psychiatry, the Hebrew University and Hadassah School of Medicine Jerusalem. Department of Psychiatry, Hadassah University Hospital, Jerusalem, Israel. Director and founder, Center for Traumatic Stress, Hadassah University Hospital, Jerusalem. INTRA- Israel National Therapeutic Riding Association.

References

Ballenger, J., Davidson, J., Lecrubier, Y., Nutt, D., Marshall, R., Nemeroff, C., Shalev, A., Yehuda, R., (2004). Consensus statement update on Posttraumatic Stress Disorder from the International Consensus Group on Depression and Anxiety, *J Clinic Psychiatry*, Vol. 65, No. 1, 55-62.

Du, S., Janniste, T., Champion, D., Yap, C., (2008). Theories of Fear Acquisition: The Development of Needle Phobia in Children, *Pediatric Pain Letter*, Vol. 10, No. 2, 13-16.

Joseph, R., (2006). Brain Mind Lecture 6, *The Limbic System, Amygdala, Hypothalamus, Hippocampus, PTSD*, http://images.mitrasites.com/limbic-system.html, last accessed the site, Dec, 23rd, 2011.

Kidd, S., & Kidd, J., (1972). Theory of Fear as a Motivation, Reprinted from *Stanislaus Journal of Psychology*, Vol. 2, No. 2, http://www.inbetweenness.com/Jim's%20P, last accessed the site, Dec, 23rd, 2011.

Milad, M., Rauch, S., Pitman, R., Quirk, G., (2006). Fear Extinction in Rats Implications for Human Brain Imaging and Anxiety Disorders, *Biological Psychology,* (Elsevier), Vol. 73, 61-71. Online Science Direct.

Neimark, N., (2011). The Flight and Fight Response, Mind /Body Education Center, http://www.thebodysoulconnection.com/EducationCenter/fight.html, last accessed the site, Dec, 23rd, 2011.

Peri, T., Ben-Shakar, G., Scott, P., Shalev, A., (2000). Psychophysiologic Assessment of Adverse Conditioning in Posttraumatic Stress Disorder, *Biological Psychiatry*, Vol. 47, 512-519.

Phelps, E., Emotions and Cognition: Insights from Studies of Human Amygdala. *Annual Review Psychology*, Vol. 57, 27- 53.

Riede, D., (1988). *Physiotherapy on Horses*, translated into English, from the original published in German in (1986) *Therapeutisches Reiten in der Krankengymnastik*. Publication in English, Omnipress Madison WI, USA, 43.

Sanders, M., Wiltgen, B., Fanslow, M., (2003). The Place of the Hippo Campus in Fear Conditioning, *European Journal of Psychology*, Vol. 463, 217-223.

Schlogmann, W. (2006). A Life Long Mathematics Learning – A Threat or an Opportunity: Some Remarks on Affective Conditions in Mathematics Courses, *Adults Learning Mathematics, An International Journal*, Vol. 2, 6-18.

Shalev, A., (2002). Acute Stress Reactions in Adults, *Society of Biological Psychiatry*, Vol. 51, 532 – 543.

Shalev, A., Bonne, O., Eth, S., (1996). Treatment of Posttraumatic Stress Disorder: A Review, *Psychosomatic Medicine,* Vol. 58, 165-182.

Shalev, A., (2008). *Living in Fear*, reported lecture on Laika's MedLiblog https://laikaspoetnik.wordpress.com/tag/shalev/, last accessed the site, Dec, 23rd, 2011.

Shkedi, A., (2011). *Therapeutic Riding and Traumatic Brain Injury*, (in press).

ARTICLE EIGHT

Hippology in Anthropology: tracking horse-human ecologies from domestication to equine-assisted therapy in anthropological discourse.[1]

Corey Souza, Ph.D. student, Dept. of Anthropology, University of Florida, USA
Correspondence to:
Corey Souza, Department of Anthropology, Turlington Hall, Room 1112 Gainesville, FL 32611-7305. University of Florida, USA
Phone: +1 352-316-0682
Fax: +1 352-392-6929
Email: csouza@ufl.edu

ABSTRACT

Research in hippology lacks a fixed location within North American academia. Crossing the boundaries of science and art, hippology has entered the discourses of archeologists, classical historians, art historians, cognition scientists, physical and psychotherapists. This paper locates the intersections of hippology and anthropology by analyzing studies ranging from early domestication to hippotherapy and proposes horse-human ecology as a dynamic field of inquiry for anthropological research.

KEYWORDS

Anthropology; Hippology; Hippotherapy; Equine Ecology; Animal Cognition.

---

[1] This paper is the result of the Spring 2011 Proseminar in biological anthropology taught by Dr. Connie Mulligan

Introduction

Hippology, the study of horses or equine science, has been addressed from numerous angles in anthropology, yet is rarely labeled as such. Straddling multiple sectors of anthropological inquiry – such as archeology, adaptation, classical history, performing arts, animal cognition, physical and psychotherapy – hippology studies are challenged by the lack of a cohesive perspective. Hippology in scholarly journals appears in research on ecological perspectives of human development, animal cognition, and alternative therapies. More difficult to find in anthropological journals, is research addressing the role of the psychological bond between equine dependent societies and horses, the evolution of communication between Equus caballus and Homo sapiens, and the possible effects of the evolution of equine science practices on equine and human cognitive capacities. Despite hippology's elusive nature, fluctuating between science and art, and the apparent waning of interest by anthropologists in the last quarter century, this paper proposes the field of hippology as a relevant mode of inquiry for contemporary anthropological research.

Methods

One of the motives in writing this paper was to ascertain the salience of hippology research in current anthropological discourse. A word search for terms: "hippology," "hippotherapy," "horse," "equine" and "equestrian," was first conducted through Anthrosource, a service of the American Anthropological Association, (http://www.anthrosource.net.lp.hscl.ufl.edu/Default.aspx) and limited to publications between 1980 and 2010. Founded in 1902, the American Anthropological Association is the leading academic reference for North American anthropologists and the largest anthropological association in the world (www.aaanet.org/about). Due to the scarcity of hippology studies located through Anthrosource, further searches were made with the above keywords and time frame in the data bases: Anthropology Plus, ISI Web of Knowledge, PLoS One and Wikipedia. These searches led to articles published in a variety of academic journals oriented towards social and behavioral science, cultural studies, linguistics, archaeology, psychology, physiotherapy and genetics research. Returns on these searches required selection of a manageable number of journals and articles for review. *Current Anthropology, American Journal of Physical Anthropology,* and *Past & Present* were selected as primary resources, in addition to *American Anthropologist*, given their status within the anthropological community. *Current Anthropology* is the primary publication of the Wenner-Gren Foundation, a private foundation dedicated exclusively to anthropological research. *The American Journal of Physical Anthropology* was chosen for its prestige in archeological studies of human evolution, including behavioral patterns. *Past & Present* is a journal geared towards history, society and culture, and is a publication of the Oxford University Press, "one of the world's oldest and largest university publishers" (www.oxfordjounrals.org/about_us.html). Journals cross-cutting cultural, linguistic, medical anthropology and archaeozoology were also referenced in order to further locate and identify the nexus of contemporary academic research pertaining to hippology and anthropology. These journals include: *Animal Cognition, Adapted Physical Activity Quarterly, Equine Veterinary Science, Physiotherapy,* and *Anthrozoös*. Searches in *Animal Cognition* and *Anthrozoös* were limited to the periods 1998-2010 and 2007-2010, respectively, due to the lack of complete archives available through the online library network at the University of Florida. Additionally, websites of the American Horse Council and the Federation of Riding for the Disabled International were consulted because each of these institutions publishes regular articles and reports on equine science, hippotherapy and the equine industry.

This paper does not claim to provide a fully comprehensive review of all articles relevant to hippology and anthropology for the following reasons: given the goal of locating hippology within current mainstream anthropology, many articles were excluded for having been published in obscure academic journals; non-

English language articles were also excluded, although it should be observed there is a wealth of hippology research taking place within European and Southwest Asian anthropological circles. Many of the articles reviewed were, in fact, studies conducted by foreign institutions and published in English. Articles appearing in search engines, but not readily available online were also left out of review. Furthermore, articles with nominal references to the above search terms (i.e., search terms appearing less than four times within an article) were removed from consideration. Lastly, articles lacking a strong correlation between horses and humans were not reviewed.

RESULTS

The initial word search through Anthrosource produced 18 results, nine of which were published in *American Anthropologist*. Of those nine results, eight consisted of book reviews and or responses to the one original paper (Osborn, 1983). Searches within auxiliary databases yielded large numbers of returns (700-4000 matches), however, once scrutinized for relevance (and publication in English), the numbers were quite small: Anthropology Plus (93), ISI Web of Knowledge (21) and Plos One (0). The task of separating relevant articles from non-relevant articles was complicated by the fact that the term "horse" is often used in metaphoric phrases: "beating a dead horse," "Trojan horse," "from the horse's mouth" and "never look a gift horse in the mouth." After completing the above searches and evaluating for relevance, articles were selected from *Current Anthropology* (14)[2] , *American Journal of Physical Anthropology* (2), *Past & Present* (4), *Animal Cognition* (12), *Anthrozoös* (2), *Equine Veterinary Journal* (8), *Physiotherapy* (1), and *Adapted Physical Activity Quarterly* (2). Furthermore, searches through Wikipedia, The American Horse Council and Federation of Riding for the Disabled International led to websites, articles and reports not found in the other databases.

Articles focused on the role of animal domestication in evolving human subsistence patterns can be found in *American Anthropologist* (Osborn, 1983) and *Current Anthropology* (Binford et al., 1986; White, 1989; Straus, 1991; Adams, 1992; Bicho, 1994; Koryakova & Kohl, 2000). The same journals present perspectives on the significance of horses in constructing cities, waging wars, building economies, marking social status and expansion of equine related leisure and art, as do articles found in *American Journal of Physical Anthropology* and *Past & Present*: (Lindner, 1981; Trinkaus, 1984; Davis, 1986; Lewis-Williams et al., 1988; Singleton, 1993; Huang, 2002; Kwint, 2002; Belcastro et al., 2007; Roche, 2008). Recent anthropological research pertaining to hippology most often pertains to archaeozoology, or the study of animal remains in archeological sites. However, the significance of equine cognition in human societies is addressed in *Animal Cognition* (McKinley & Sambrook, 2003; Kreuger & Heinze, 2008; Maros et al., 2008; Lemasson et al., 2009; Uller & Lewis, 2009; Farmer et al., 2010; Hothersall et al., 2010; Proops & McComb, 2010; Sankey et al., 2010). Equine psychology and horse-human ecologies are also placed under the microscope in *Equine Veterinary Journal* (Cooper, 1998; Goodwin, 1999; Robinson, 1999). Web-based resources Wikipedia, American Horse Council and Federation of Riding for the Disabled International also offer useful bibliographies and links for researchers interested in multifaceted nature of equine ecologies.

DISCUSSION

Finding information pertinent to both anthropology and hippology is challenged by the lack of a cohesive aggregate of social science researchers working in this field (Edenburgh, 1999; Digard, 2008). Roche

---

[2] The search within Current Anthropology led to 371 results. Of those articles, the first 100 (as presented by the search engine in order of relevance to search inquiry) were scanned for frequency of search terms. Of those 100 reviewed, 14 articles were found to be relevant to this paper.

(2008) describes written documentation in hippology as that which was created by and for elite members of society familiar with the lexicon pertaining to horsemanship. The majority of knowledge in equine science and art however, is found in corporeal culture, that is, performance-based expertise passed down from one master equestrian to the next. The challenge in tracing hippology's history is: *"How to understand the practices of the real through discursive practices, that is to say, to summarize Michel Foucault's definition, the set of rules which are immanent to a practice and which define its specificity — which comes down to relating words to things, and experiences to discourses"*. (Roche, 2008: 130).

The challenge of translating "experiences to discourses" is what all anthropologists face when recording human phenomena for scientific purposes. Nonetheless, it is an endeavor allowing for particular sets of cultural information to become globally digestible. In order to supplement the lack of academic texts in equestrian history, Roche turns to financial and breeding documents, literary accounts, technical manuals, and visual documents such as paintings, pictures, statues and cinema.

An anthropological approach to hippology research would amplify the knowledge base in the fields of medical anthropology and hippotherapy, a form of therapy claimed to improve motor-sensory function performance for patients suffering from brain damage, multiple sclerosis, and cerebral palsy, among other ailments (DePauw, 1986). Often considered 'alternative therapy', hippotherapy has implications for both physical and psychological treatments (Debuse et al., 2005; Liddiard, 2008; Yorke et al., 2008; Selby, 2009), yielding promising results in ameliorating the adaptive capacities of autistic individuals, although the methods and results of these studies have been contested. Pauw suggests larger sample sizes in order to clearly distinguish "between a statistically significant effect and a clinically meaningful effect" (Pauw, 2000: 525). Researchers can turn to specialized journals such as *Proceedings of the International Therapeutic Riding Congress* and *Scientific and Educational Journal of Therapeutic Riding*, which are rich in research on the merits of equine-assisted therapy. Hippotherapy has also made appearances in *Physiotherapy* and *Adapted Physical Activity Quarterly*, among other health related journals.

Despite the unique interdisciplinary nature of hippotherapy, this topic is almost invisible on the anthropological radar. Horses' roles in human society have only recently seen a transition from primary to periphery. A bio-cultural-ecological approach to hippology is a promising resource for understanding the human psyche and improved theory and method in rehabilitative psycho-social therapies.

EQUINE DOMESTICATION: A REVOLUTIONARY CHANGE IN HUMAN BEHAVIOR

Hippology in anthropology is not uncommon in research on domestication and subsequent effects on human subsistence and migration patterns. Horse domestication is often subsumed by the broad-spectrum analysis of domestication of plants and animals. The fundamental change from hunting and gathering to plant and animal husbandry constituted both a biological and a cultural revolution, although the exact dates and locations of diverse aspects of this revolution are still contested amongst archaeologists (White, 1989; Levine, 1999; Erickson et al., 2000; Koryakova & Kohl, 2000; Lamberg-Karlovsky, 2002). The extraordinary effects that horse domestication had on human populations due to their versatility as food, transport and aesthetic (never mind therapeutic) members of human ecologies has been dealt with in anthropological, archeological and scientific journals, but finding those articles requires knowing when, where and how equines have been relevant to human society. Boone and Smith (1998) discuss the primacy of natural selection versus adaptive decision-making in the process of animal domestication, while Shipman (2010) identifies the human-animal relationship as one of the key factors in the unique process of human evolution.

A word search for instance of "hippology" or "hippotherapy" yields no results in *American Anthropology, Current Anthropology,* or *American Journal of Physical Anthropology*. When searching the terms "horse," "equine" and "equestrian," these journals yield increasingly higher volumes of scholarly articles relevant to these topics. Upon close examination, however, one finds numerous book reviews and articles with nominal references to horses and equestrian culture.

Of the relatively few articles focused on hippology studies found in these journals, the majority appear in the early to mid-twentieth century and focus on the effects of horse domestication on (semi)nomadic cultures. According to Alvard and Kuznar (2001), the transition from hunting and gathering subsistence methods to food production resulted in sedentism, population growth and increased socio-political complexity in the late Pleistocene-early Holocene epochs. These authors question the established assumption of animal domestication as a direct response to reduced resources and highlight a fundamental transition in short-term to long-term planning on the part of human populations (Alvard & Kuznar, 2001).

Horse domestication has been located between five and eight thousand years ago in the modern day region of Kazakhstan. A topic of contestation in determining the earliest date of domestication lies in delineating the definition of domestication. Control of breeding for food purposes certainly predated the use of equines in transportation and warfare. Furthermore, understanding the needs of equine subsistence has brought previously held assumptions about equestrian cultures under scrutiny (Lindner, 1981; Osborn, 1983).

Analyzing historical documents and archeological records pertaining to the Huns' dominance of Eurasia circa 2 kya, Lindner (1981) questions the validity of labeling the Huns as an equestrian culture post-expansion westward of the Carpathian Mountains. Known by their rivals and admirers as warring centaurs, the Huns dominated territories north of the Black Sea from the backs of horses. It is doubtful, however, that the Huns remained mounted after traversing the Carpathians, due to environmental factors that would not have been conducive to maintaining the health and wellbeing of horses fit for battle. This climatic change would yield a fundamental transition for the Hunnic way of life and ultimately, this "loss undermined their military strength" (Lindner, 1981: 17).

In a call for increased attention to ecological and ethno-historical aspects of anthropology, Osborn suggests that the introduction of horses to North American Blackfoot, Shoshoni-Comanche, Apache and Kiowa cultures had a profound effect on the technology, subsistence, mobility, development, warfare, and socialization of these groups (Osborn, 1983). Acknowledging prior literature on the topic describing Native American tribes as pre-adapted to an equestrian lifestyle, Osborn offers a view of the **fundamental adaptations** made by native North Americans in their adoption of equine culture. With systematic analysis of the horse-human ratio in Western North America, Osborn seeks heightened understanding of the behavioral adaptations of Native American cultures in light of the introduction of horses to their way of life. In order to do this, anthropologists must "*better understand the nature of environmental stress on horse populations[.]*" Only then will they "*be better equipped to investigate associated adaptive responses to equestrian groups*" (Osborn 1983: 583). Levine (1999) also stresses the importance of understanding the history of horse husbandry in order to better understand our own past.

These authors' valuable insights into the need for heightened understanding of equine subsistence patterns in order to paint a clearer picture of human evolution from hunting and gathering to equestrian cultures, were among few pertaining to hippology studies within their respective journals.

Taking a brief glance at the above search terms through Wikipedia (2011; 2011a; 2011b), one encounters a valuable bibliography relevant to specific effects of horse domestication on ancient and Native American cultures in a variety of scholarly journals such as: *Journal of Indo-European Studies, Antiquity, Journal of Archeological Science, Scientific American, Molecular Ecology, Molecular Biology and Evolution, Nature Genetics, Science, South Asian Studies* and *Publications of the McDonald Institute for Archeological Research*. Of the 56 books and journal articles listed in Wikipedia's bibliography, only three were published prior to the year 1990, while the majority was published in the 21st century.

The search for "hippotherapy" in the Wikipedia database guides inquiries to websites of the American Hippotherapy Association, the Federation of Riding for the Disabled International and North American Riding for the Handicapped. Despite Wikipedia's lack of academic prestige, it is a useful tool for finding resources on a given topic. It is encouraging to find a consolidated resource for recent publications pertaining to hippology.

Upon closer examination of the Wikipedia's equine bibliography, two authors appear more frequently than others: Marsha Levine and David Anthony. Dr. Levine is a senior research associate at the McDonald Institute for Archeological Research at Cambridge. Her research interests include: the transition from horse hunting/herding to horse husbandry and the development for elucidation of the horse-human relationship, and ethnoarcheology. It is these last areas of research that distinguish Dr. Levine's approach from traditional inquiries into the nature of the role of equines in human society. Dr. Levine's most recent project *From Chariot to Equestrian Pastoral Nomadism: the Evolving Role of the Horse*, was supported by the Leverhulme Trust (2007-2010). Dr. Levine and her collaborators have published ten articles through the MacDonald Institute and in numerous journals such as *Equine Veterinary Science* and *the Journal for Anthropological Archeology* (Levine, 2010).

In the United States, Dr. David Anthony worked in conjunction with his wife Dorcas Brown to found the Institute for Ancient Equestrian Studies (IAES) at Hartwick in 1994 where they continue to serve as institute directors. IAES has received funding from the American Philosophical Society, Wenner-Gren Foundation, the National Science Foundation, Freedman Foundation and Hartwick College. Dr. Anthony and Ms. Brown have co-authored several hippology articles, which can be found in diverse publications such as: Current Anthropology, Scientific American, and Antiquity (Anthony & Brown, 2011).

Equine Cognition: Do Horses Know Who We Are?

Research in cognitive science has focused on primates for two reasons: their genetic similarity to human beings and the legal/ethical difficulties in conducting cognition experiments with human subjects. However, recent studies in *Animal Cognition, Anthrozoös* and *Equine Veterinary Journal* reveal the significance of inquiries into the cognition of non-primates as well. Recent studies in equine cognition demonstrate that human beings play a significant role in the minds of domestic horses. Furthermore, the study of equine social patterns presents learning opportunities for human societies. In the last ten years alone, *Animal Cognition* has published nine articles specifically related to research in equine cognition, proving that horses are capable of long-term memory, learning and other cognitive tasks previously assumed to be the privileged arena of primates.

A popular test for evaluating the capacity of equine cognition has been the use of pointing gestures, an experiment previously used with non-human primates and canines (McKinley & Sambrook, 2003; Maros et al., 2008; Proops & McComb, 2010). Equines have presented mixed results in their capacity to properly identify human intentions with pointing gestures, however, these experiments have led to the discovery of horses' bias towards spatial memory (Hothersall et al., 2010). Despite mixed results in response to pointing gestures, it is clear that horses have developed a sophisticated capacity to read human body language, that is, kinesthetic and postural indicators of mood and intention (Proops & McComb, 2010). One complicating factor in unsuccessful experiments with pointing gestures is that horses are sensitive to members of their own social groups (Kreuger & Heinze, 2008) and may not be interested in the pointing gestures of a researcher who has not established a social bond with the test horse. Even when experiments are conducted using the horse's regular handler or owner, the horse's response may vary depending upon the type of training the horse has received throughout his or her life (McKinley & Sambrook, 2003).

Employing non-invasive research methods, Hanggi and Ingersoll (2009) report on long-term memory experiments in equine categorization capacities revealing an adaptive ability to generalize over diverse testing conditions. Stone (2009) confirms evidence of horses' ability to recognize individual humans based on two-dimensional images and Fureix et al. (2009) demonstrate that human beings represent relevant entities in equine ecology.

Hanggi and Ingersoll tested the memory of three horses' performance of cognitive tasks: discrimination learning, categorization and concept use, over six to ten year intervals. The first experiments conducted in 1997 and 2001 established that horses are capable of object association, categorization and relative size conceptualization. Testing for object association, Hanggi and Ingersoll found that horses were better able to remember object associations found at the beginning and end of arbitrary lists, which supports principles of "primacy and/or recency effects" in learning (Hanggi & Ingersoll, 2009: 455). Their second test involved a trial and error period of learning. Once the horses had learned the discriminations required for the test, they scored accurately on new items presented for discrimination, i.e. "learning to learn" (Hanggi & Ingersoll, 2009: 456). The last test required horses to not only evaluate relative size concepts, but also to generalize between two and three-dimensional stimuli. Hanggi and Ingersoll conclude that further understanding of equine cognition will assist training and management of equines.

Stone (2010) conducted a study using operant conditioning to determine whether horses could distinguish faces from photographs. Her results indicate that horses learned to discriminate photographs of unrelated individuals. Stone relates her work to future implications in hippotherapy, which has been shown to ease suffering in patients with physical and emotional disorders: depression, ADD, ADHD and PTSD (DePauw, 1986; Yorke et al., 2008; Selby, 2009; Engel, 1994 cited in Stone, 2010). Stone's work suggests that facial recognition is not a behavioral or cognitive process, but a combination of both. Stimulus and recognition work in concert with one another. This information can assist in hippotherapy programs by shortening the time needed to acquaint horse and rider by hanging a photograph of the patient in the horse's stall (Stone, 2010).

Cooper (1998), Lemasson et al. (2009), Uller and Lewis (2009), Farmer et al. (2010) and Sankey et al. (2010) have contributed studies to the heightened understanding of equine learning processes. This research can improve horse-human relationships, such that horse handlers may have more rewarding experiences interacting with equines, and horses may not fall victim to euthanization due to behavioral problems (Warren-Smith & McGreevy, 2008). Goodwin illuminates the imposition of human cultural biases in our interactions with equines, preferring dominance over cooperation. Studies of feral equine societies show that horses are cooperative by nature and avoid conflict as a means for survival (Goodwin, 1999).

Conclusion

Hippology in Anthropology: How Relevant Is It?

According to the American Horse Council, a study published in 2005 reveals that the horse industry constitutes *"$39 billion in direct economic impact to the US economy and supports 1.4 million jobs on a full-time basis."* Evaluating indirect economic impacts, the same study estimates a contribution of $102 billion to the United States economy and counts 9.2 million horses in the U.S. alone. This information reveals that horses and equestrian culture constitute a significant sector of North American cultural and economic practices. The horse industry is on par with the American Brewing Industry, which employs 1.9 million Americans and contributes $103 billion in direct and indirect impacts on the U.S. economy (Beer Serves America, 2011).

From a biological perspective, horses stand to offer humans much more than improved subsistence methods. Founded in 1980, The Federation of Riding for the Disabled International has been publishing research on the merits of hippotherapy for over fifteen years and sponsors the International Congress of

Therapeutic Riding. Their mission is to *"facilitate the worldwide collaboration between organizations and individuals whose objectives are philanthropic, scientific and educational in the field of equine-assisted activities"* (Federation of Riding for the Disabled International, 2010). Goodwin (1999), Robinson (1999) and York et al. (2008) emphasize the potential for improving human social development with lessons learned through equine ecology. Breen (2002) and Ghio et al. (2006) explore the mutual benefits of cytogenetic and medical research for humans and horses.

Although hippology continues to appear in discussions on human subsistence patterns, this area of research has taken a back seat in North American anthropology since the mid-twentieth century, save for a few specialized research institutions. Hippotherapy is even further from the main stream, making its way into scientific journals pertaining to physical therapy and cognition. Given the multifaceted contributions equines have made to human societies, hippology remains an untapped resource in anthropological research. Further inquiry from biological, cultural, psychological and ecological perspectives, stands to make a significant contribution to knowledge production in the fields of anthropology and hippology alike.

Acknowledgements: I would like to thank the Department of Anthropology at the University of Florida, my advisor Dr. Faye Harrison and committee members: Dr. Michael Heckenberger, Dr. Florence Babb and Dr. Joan Frosch. I would also like to thank Dr. Connie Mulligan, whose course in biological anthropology was the inspiration for this paper.

References

Adams, R. (1992). Anthropological Perspectives on Ancient Trade. *Current Anthropology*. 33(1): 141-160.

Alvard, M. and Kuznar, L. (2001). Deferred Harvests: the Transition from Hunting to Animal Husbandry. *American Anthropologist*. 103(2): 295 – 311.

American Horse Council (2005). The Economic Impact of the Horse Industry on the United States National and State Breakout Report [Internet]. Washington, DC (United States): American Horse Council. [cited February 2011]. Retrieved from: http://www.horsecouncil.org/product/2005-economic-impact-horse-industry-state-breakout-reports.

Anthony, D. (1986). The "Kurgan Culture," Indo-European Origins, and the Domestication of the Horse: A Reconsideration. *Current Anthropology*. 27(4): 291 – 313.

Anthony, D. and Brown, D. (2011). Eurasian Steppe Archeological Research: Institute for Ancient Equestrian Studies [Internet]. United States: Hartwick College. [cited February 2011]. Retrieved from: http://users.hartwick.edu/anthonyd/home.html.

Beer Serves America. (2011). Economic Impact [Internet]. United States: Beer Serves America. [cited February 2011]. Retrieved from: http://www.beerservesamerica.org/economic/default.aspx.

Belcastro, G.; Rastelli, E.; Mariotti, V.; Consiglio, C.; Facchini, F.; Bonfiglioli, B. (2007). Continuity or Discontinuity of the life-style in Central Italy During the Roman Imperial Age-Early Middle Age Transition: Health, Diet and Behavior. *American Journal of Physical Anthropology*. 132: 881-394.

Bicho, N.F. (1994). The End of the Paleolithic and the Mesolithic in Portugal. *Current Anthropology*. 35(5): 664-674.

Binford, L,; Stone, N; Aigner, J.; Behrensmeyer, A.; Haynes, G.; Olsen, J.; Todd, L.; Yu-Zhu, You. (1986). Zhoukoudian: A Closer Look. *Current Anthropology*. 27(5): 453-475.

Boone, J. and Smith, E. (1998). Is It Evolution Yet? A Critique of Evolutionary Archaeology. *Current Anthropology*. 39(S1): S141-S174.

Breen, M. (2002). Equine Clinical Cytogenetics – Human Chromosomes Sitting on Horse Chromosomes. *Equine Veterinary Journal*. 34(2): 110-111.

Cooper, J.J. (1998). Comparing Learning Theory and its Application in the Training of Horses. *Equine Veterinary Journal*. S27: 39-43.

Davis, W. (1986). The Origins of Image Making. *Current Anthropology*. 27(3): 193-215.

Debuse, D.; Chandler, C.; and Gibb, C. (2005). An Exploration of German and British Physiotherapists' Views on the Effects of Hippotherapy and Their Measurement. *Physiotherapy Theory and Practice*. 21(4): 219-242.

DePauw, K. (1986). Horseback Riding for Individuals with Disabilities: Programs, Philosophy and Research. *Adapted Physical Activity Quarterly*. 3: 217-226.

Digard, J.P. (2008). Waltham Symposium Abstracts: Research in the Social Science of Horses: why and how? *Equine Veterinary Journal*. 40(6): 56-57.

Edenburgh, N. (1999). Perceptions and Attitudes Towards Horses in European Societies. *Equine Veterinary Journal*. S28: 38-41.

Erickson, J.D., Lee, D. and Bertram, J.E.A. (2000). Fourier Analysis of Acetabular Shape in Native American Arikara Populations Before and After Acquisition of Horses. *American Journal of Physical Anthropology*. 113: 473-480.

Farmer, K., Kreuger, K., and Byrne, R. (2010). Visual Laterality in the Domestic Horse (Equus caballus) Interacting with Humans. *Animal Cognition*. 13: 229-238.

Federation of Riding for the Disabled International. (2010). Publications [Internet]. Maine (United States): Federation of Riding for the Disabled International. [cited February 2011]. Retrieved from: http://www.frdi.net/publications.html.

Fureix, C., Jego, P., Sankey, C., Hausberger, M. (2009). How Horses (Equus caballus) See the World: humans as significant "objects". *Animal Cognition*. 12: 643 – 654.

Ghio, A.J., Mazan, M.R., Hoffman, A.M., Robinson, N.E.(2006). Correlates Between Human Lung Injury After Particle Exposure and Recurrent Airway Obstruction in the Horse. *Equine Veterinary Journal*. 38(4): 362-367.

Goodwin, D. (1999). The Importance of Ethology in Understanding the Behavior of the Horse. *Equine Veterinary Journal*. S28: 15-19.

Hanggi, E and Ingersoll, J. (2009). Long-term memory for categories and concepts in horses (Equus caballus). *Animal Cognition*. 12: 451-462.

Hothersall, B., Gale, E.V. and Nicol, C.J. (2010). Cue Use by Foals (Equus caballus) in a Discrimination Learning Task. *Animal Cognition*. 13: 63-74.

Huang, H.T. (2002). Hypoclastia and the Chinese Diet. *Current Anthropology*. 43(5): 809-819.

Koryakova, L. and Kohl, P. (2000). Complex Societies From the 3rd to the 1st Millennia b.c.: Regional Specifics in the Light of Global Models. *Current Anthropology*. 41(4): 638-642.

Kreuger, K. and Heinze, J. (2008). Horse Sense: social status of horses (Equus caballus) affects their likelihood of copying other horses' behavior. *Animal Cognition*. 11: 431-439.

Kwint, M. (2002). The Circus in Late Georgian England. *Past & Present*. 174: 73-115.

Lamberg-Karlovsky, C.C. (2002). Archeology and Language: The Indo-Iranians. *Current Anthropology*. 43(1): 63-88.

Lemasson, A., Boutin, A., Boivin, S., Blois-Heulin, C., Hausberger, M. (2009). Horse (Equus caballus) Whinnies: a source of social formation. *Animal Cognition*. 12: 693-704.

Levine, M. (1999). Investigating the Origins of Horse Domestication. *Equine Veterinary Journal*. S28: 6-14.

Levine, M. (2010). *University of Cambridge Department of Archeology: Marsha Ann Levine* [Internet]. Cambridge (United Kingdom): University of Cambridge. [cited February 2011]. Retrieved from: http://www.arch.cam.ac.uk/~ml12/.

Lewis-Williams, J.D., Dowson, T.A., Bahn, P., Bandi, H.G., Bednarik, R., Clegg, J., Consens, M., Davis, W., Delluc, B., Delluc, G., Faulstich, P., Halverson, J., Layton, R., Martindale, C., Mirimanov, V., Turner II, C., Vastokas, J., Winkelman, M., Wylie, A. (1988). The Signs of All Times: Entopic Phenomena in Upper Paleolithic Art. *Current Anthropology*. 29(2): 201-245.

Liddiard, J. (2008). The Use of Hippotherapy to Support Function and Participation in Children with Disabilities. *Journal of Intellectual and Developmental Disability*. 33(3): 281.

Lindner, R. P. Nomadism, Horses, and Huns. (1981). *Past & Present*. 92: 3-19.

Maros, K., Gácsi, M. and Miklósi, Á. (2008). Comprehension of Human Pointing Gestures in Horses (Equus caballus). *Animal Cognition*. 11: 457-466.

McKinley, J. and Sambrook, T. (2003). Use of Human-Given Cues by Domestic Dogs (Canis familiaris) and Horses (Equus caballus). *Animal Cognition*. 3: 13-22.

Osborn, A. (1983). Ecological Aspects of equestrian Adaptations in Aboriginal North America. *American Anthropologist*. 85: 563 – 591.

Pauw, J. (2000). Therapeutic Horseback Riding Studies: Problems Experienced by Researchers. *Physiotherapy*. 86(10): 523-527.

Proops, L. and McComb, K. (2010). Attributing Attention: the use of human-given cues by domestic horses (Equus caballus). *Animal Cognition*. 13:197-205.

Robinson, I.H. (1999). The Human-Horse Relationship: how much do we know? *Equine Veterinary Journal*. S28: 42-45.

Roche, D. (2008). Equestrian Culture in France from the Sixteenth to the Nineteenth Century. *Past & Present*. 199: 113 – 145.

Sankey, C., Richard-Yris, M., Henry, S., Fureix, C., Nassur, F., Hausberger, M. (2010). Reinforcement as a Mediator of the Perception of Humans by Horses. *Animal Cognition*. 13:753-764.

Selby, A. (2009). A Systematic Review of the Effects of Psychotherapy Involving Equines. (M.S. Thesis). Arlington: University of Texas. Retrieved from Google Scholar https://dspace.uta.edu/handle/10106/1695.

Singleton, J. (1993). Britain's Military Use of Horses 1914-1918. *Past & Present*. 139: 178-203.

Shipman, P. (2010). The Animal Connection and Human Evolution. *Current Anthropology*. 51(4): 519-538.

Stone, S. (2010). Human facial discrimination in horses: can they tell us apart? *Animal Cognition*. 13: 51-61.

Straus, L. W. (1991). Southwestern Europe at the Last Glacial Maximum. *Current Anthropology*. 32(2): 189-199.

Trinkaus, K.M. (1984). Mortuary Ritual and Mortuary Remains. *Current Anthropology*. 25(5): 674-679.

Uller, C. and Lewis, J. (2009). Horses (Equus caballus) select the greater of two quantities in small numerical contrasts. *Animal Cognition*. 12: 733-738.

Warren-Smith, A. and McGreevy, P. (2008). Equestrian Coaches' Understanding and Application of Learning Theory in Horse Training. *Anthrozoös*. 21(2): 153-162.

White, R. (1989). Husbandry and Herd Control in the Upper Paleolithic: A Critical Review of the Evidence. *Current Anthropology*. 30(5): 609-632.

Wikipedia. (2011). Domestication of the Horse [Internet]. United States: Wikimedia Foundation. [cited February 2011]. Retrieved from: http://en.wikipedia.org/wiki/Horse_domestication

Wikipedia. (2011a). Hippology [Internet]. United States: Wikimedia Foundation. [cited February 2011]. Retrieved from: http://en.wikipedia.org/wiki/Hippology.

Wikipedia. (2011b). *Hippotherapy* [Internet]. United States: Wikimedia Foundation. [cited February 2011]. Retrieved from: http://en.wikipedia.org/wiki/Hippotherapy.

Yorke, J., Adams, C. and Coady, N. (2008). Therapeutic Value of Equine-Human Bonding in Recovery from Trauma. *Anthrozoös*. 21(1): 17-30.

Made in United States
Orlando, FL
04 November 2026

90206126R00050